ROBERT JAMESON

MINERALOGY

ACCORDING TO THE

NATURAL HISTORY SYSTEM

FORMING THE ARTICLE UNDER THAT HEAD IN THE

SEVENTH EDITION OF THE

ENCYCLOPAEDIA BRITANNICA

Elibron Classics
www.elibron.com

Elibron Classics series.

ISBN 1-4021-6407-6 (paperback)
ISBN 1-4021-0348-4 (hardcover)

This Elibron Classics Replica Edition is an unabridged facsimile of the edition published in 1837 by Adam and Charles Black, Edinburgh.

MINERALOGY

ACCORDING TO THE

NATURAL HISTORY SYSTEM.

PRINTED BY NEILL AND CO., OLD FISHMARKET,
EDINBURGH.

MINERALOGY

ACCORDING TO THE

NATURAL HISTORY SYSTEM,

FORMING THE ARTICLE UNDER THAT HEAD IN THE
SEVENTH EDITION OF THE

ENCYCLOPÆDIA BRITANNICA.

BY

ROBERT JAMESON, F.R.SS.L. & E., M.W.S.

REGIUS PROFESSOR OF NATURAL HISTORY, AND LECTURER ON
MINERALOGY IN THE UNIVERSITY OF EDINBURGH,
&c. &c. &c.

EDINBURGH:

ADAM AND CHARLES BLACK, NORTH BRIDGE,
BOOKSELLERS TO HIS MAJESTY FOR SCOTLAND.

MDCCCXXXVII.

ADVERTISEMENT.

THIS short Treatise, originally printed in the Supplement to the Encyclopædia Britannica, and again in the New Edition of the same national work, contains, besides an explanation of the Natural History System of Mineralogy, also descriptions of the more important mineral species. Although published in its present form chiefly for the use of Students of Mineralogy, it may be consulted by the more advanced Mineralogist.

The Natural History System of Mineralogy was first made known to British mineralogists in the third edition (in three volumes) of my System of Mineralogy, published in 1820. In the year 1821, in my Manual of Mineralogy, the same method was more fully explained; and again, in the Encyclopædia Britannica, in 1822; in 1825, Mr Haidinger published a translation of Professor Mohs' " Treatise on Mineralogy, or the Natural History of the Mineral Kingdom; and, in 1834, my Manual of Mineralogy having been long out of print, Mr Allan published "A Manual of Mineralogy, according to the Natural History System, comprehending the more recent discoveries in the Mineral Kingdom."

College,
Edinburgh, 24*th May* 1837.

CORRIGENDA.

At page 98, Genus xiii. to be deleted.
— 135. in Order II. Baryte, *for* If H = 5.0, and sp. gr. less than 4.5, the streak is white. *read* If H = 5.0, the sp. gr. is less than 4.5, and the streak white
— 162, *for* Algalmatolite. *read* Agalmatolite.
— 165, in the Order Spar, *for* If the hardness = 6.0, *read* If the hardness above 6.0,
— 190, in line 10, *after* oxide of copper *add* or phosphat of iron.

CONTENTS.

CHARACTERS

OF THE

CLASSES AND ORDERS.

CHARACTERS OF THE CLASSES.

CLASS I.

GASES—WATERS—SALTS.

Specific gravity under 3.8.
If solid is sapid.
No bituminous smell.

CLASS II.

EARTHY AND METALLIFEROUS MINERALS.

Specific gravity above 1.8.
Tasteless.

CLASS III.

INFLAMMABLE MINERALS.

Specific gravity under 1.8.
If liquid, the smell is bituminous.
If solid, is tasteless.

CHARACTERS OF THE ORDERS.

Characters of the Orders of Class I.

Order I. Gas.

Specific gravity = 0.0001...0.0014.
Expansible.
Not acid.

Order II. Water.

Liquid.
Tasteless, or with sensible taste and smell.
Specific gravity = 1.0—1.0269.

Order III. Acid.

Specific gravity = 0.0015—3.7.
Acid.

Order IV. Salt.

Specific gravity = 1.2—2.9.
Solid.
Not acid.

Characters of the Orders of Class II.

Order I. Haloide.

Not metallic.
Streak white.
Hardness = 1.5 ... 5.0.
Specific gravity = 2.2 ... 3.3.

If pyramidal or prismatic, the hardness = 4.0 and less; cleavage imperfect, in oblique directions.

If tessular, the hardness = 4.0.

If single highly perfect faces of cleavage, the specific gravity = 2.4, and less.

If hardness under 2.5, the specific gravity = 2.4, and less; no resinous lustre.

If specific gravity = 2.4, and less, the hardness is under 2.5; no resinous lustre.

Order II. Baryte.

Not metallic.

No metallic-pearly lustre.

Streak, white, or orange-yellow.

Hardness = 2.5 ... 5.0.

Specific gravity = 3.3 ... 7.3.

If cleavage monotomous, the specific gravity = 4.0 and less, or = 5.0 and more. In thin folia, not elastic, not flexible.

If lustre adamantine or imperfect metallic, the specific gravity = 5.0 and more.

If Streak orange-yellow, the specific gravity = 6.0 and more.

If hardness = 5.0, the specific gravity is less than 4.5, and the streak white.

If specific gravity less than 4.0, and the hardness = 5.0, the cleavage is diprismatic.

Order III. Kerate.

Not metallic.

Streak white.

Cleavage not monotomous; not perfect peritomous.

Hardness = 1.0, 2.0.

Specific gravity 5.5 ... 6.5.

Order IV. Malachite.

Not metallic.

Colours blue, green, brown.

Hardness = 2.0 ... 5.0.

Specific gravity = 2.0 ... 4.6.

If cleavage monotomous, the specific gravity = 4.0 and more.

If colour or streak brown, the hardness = 3.0 and less; specific gravity above 2.5.

If streak blue, the hardness = 4.0 and less.

If streak white, the specific gravity = 2.2 and less; hardness = under 3.0.

Order V. Mica.

Cleavage monotomous, very distinct.

Hardness = 1.0 ... 4.5.

Specific gravity = 1.8 ... 3.4.

If metallic, the specific gravity under 2.2.

If non-metallic, the specific gravity above 2.2.

If streak yellow, the specific gravity under 3.2.

If hardness = 3.0 and more, it is rhombohedral.

If specific gravity under 2.5. it is metallic.

Order VI. Spar.

Not metallic.

Streak, white, brown, blue.

Hardness = 3.5 ... 7.0.

Specific gravity 2.0 ... 3.7.

If tessular, the specific gravity = 3.0 and less.

If rhombohedral, the specific gravity = 2.2 and less, or hardness = 6.0.

If hardness = 4.0 and less, it is rhombohedral, or cleavage monotomous, and very distinct.

If hardness above 6.0, the lustre is pearly, the cleavage perfect, and specific gravity less than 2.5, or more than 2.8.

If specific gravity above 3.3, the combination is hemi- or tetarto-prismatic; or the hardness = 6.0; no adamantine lustre.

If the specific gravity = 2.4 and less, it is not without traces of form and cleavage.

Order VII. Gem.

Not metallic.

No metallo-adamantine lustre.

Streak white or grey.

Hardness = 5.5...10.0.

Specific gravity = 1.9...4.7.

If hardness = 6.0 and less, it is tessular, and specific gravity = 3.1, and more; or specific gravity = 2.4 and less, and no traces of form and cleavage.

If specific gravity less than 3.8, there is no pearly lustre.

Order VIII. Ore.

Streak not green.

Hardness = 2.0—7.0.

Specific gravity 3.4...7.4.

If metallic, the colour is black.

If non-metallic, the lustre is adamantine or imperfect metallic.

If streak yellow or red, the hardness = 3.5 and more; specific gravity 4.8 and more.

If streak brown or black, the hardness = 5.0 and more, or 2.5 and less; or single distinct cleavage.

If hardness is 2.5 and less, the specific gravity is under 5.0.

If hardness = 4.5 and less, the streak is yellow, red, or black.

If hardness equal to 6.5 or more, and streak white, the specific gravity = 6.5 and more.

Order IX. Native Metal.

Metallic.

Colour not black.

Hardness = 0.0...5.0

Specific gravity = 5.7...20.0

If colour grey, it is malleable, and specific gravity = 7.4 and more.

If hardness above 4.0, is malleable.

Order X. Pyrites.

Metallic.

Streak black.

Hardness = 3.0...6.5.

Specific gravity = 4.1...7.7.

If hardness = 4.5 and less, the specific gravity is under 5.3.

If specific gravity = 5.3 and less, the colour is yellow or red.

Order XI. Glance.

Metallic.

Streak and colour grey and black.

Hardness = 1.0...4.0.

Specific gravity = 4.2...7.6.

If cleavage monotomous, and the specific gravity under 5.0, the colour is lead-grey.

If hardness = 2.5 and less, the specific gravity is above 5.0, or colour lead-grey.

If specific gravity above 7.4, the colour is lead-grey.

Order XII. Blende.

Streak green, red, brown, white.

Hardness = 1.0...4.0.

Specific gravity = 3.9...8.2.

If metallic, the colour is black.

If non-metallic, the lustre is adamantine.

If streak green, the colour is iron-black.

If streak brown or white, the specific gravity is between 4.0 and 4.2; form tessular, and cleavage perfect.

If streak red, the specific gravity = 4.5 and more; hardness = 2.5 and less.

If specific gravity = 4.3 and more, the streak is red.

Order III. Sulphur.

Not metallic.

Colour yellow, red, brown.

Prismatic.
Hardness = 1.0...2.5.
Specific gravity = 1,9...3.6

If cleavage monotomous, the specific gravity = 3.4 and more.

If specific gravity above 2.1, the streak is yellow or red.

Characters of the Orders of Class III.

Order I. Resin.

Hardness = 0.0...2.5.
Specific gravity = 0.7...1.6

If specific gravity = 1.2 and more, the streak is white.

Order II. Coal.

Streak brown, black.
Hardness = 1.0...2.5.
Specific gravity = 1.2...1.6.

TABULAR VIEW

OF THE

CLASSES, ORDERS, GENERA, AND SPECIES.

CLASS I.

Order I.—GAS.

Genus I. Hydrogen Gas.—1. Pure Hydrogen Gas. 2. Empyreumatic or Carburetted Hydrogen Gas. 3. Sulphuretted Hydrogen Gas. 4. Phosphuretted Hydrogen Gas.

Genus II. Atmospheric Air.—1. Pure Atmospheric Air.

Order II.—WATER.

Genus I. Atmospheric Water.—1. Pure Atmospheric Water.

Genus II. Sea Water.—1. Common Sea Water.

Order III.—ACID.

Genus I. Carbonic Acid.—1. Aeriform Carbonic Acid.

Genus II. Muriatic Acid.—1. Aeriform Muriatic Acid.

Genus III. Sulphuric Acid.—1. Aeriform Sulphuric Acid. 2. Liquid Sulphuric Acid.

Genus IV. Boracic Acid.—1. Prismatic Boracic Acid.

Genus V. Arsenic Acid.—1. Octahedral Arsenic Acid.

Order IV.—SALT.

Genus I. Natron.—1. Hemiprismatic Natron.

Genus II. Glauber Salt.—1. Prismatic Glauber Salt.

Genus III. Nitre.—1. Prismatic Nitre.

Genus IV. Rock-Salt.—1. Hexahedral Rock-Salt.

Genus V. Sal-Almoniac.—1. Octahedral Sal-Almoniac.

Genus VI. Vitriol.—1. Green Vitriol, or Sulphate of Iron; Hemiprismatic Vitriol. 2. Blue Vitriol, or Sulphate of Copper; Tetarto-prismatic Vitriol. 3. White Vitriol, or Sulphate of Zinc; Prismatic Vitriol.

Genus VII. Epsom Salt.—1. Prismatic Epsom Salt.

Genus VIII. Alum.—1. Octahedral Alum.

Genus IX. Borax.—1. Prismatic Borax.

Genus X. Glauberite.—1. Prismatic Glauberite.

CLASS II.

Order I.—HALOIDE.

Genus I. Gypsum.—1. Prismatoidal Gypsum, or Common Gypsum. 2. Prismatic Gypsum, or Anhydrite.

Genus II. Cryolite.—1. Prismatic Cryolite.

Genus III. Alum Stone.—1. Rhombohedral Alum Stone.

Genus IV. Fluor.—1. Octahedral Fluor.

Genus V. Apatite.—1. Rhombohedral Apatite.

Genus VI. Limestone.—1. Aragonite, or Prismatic Limestone. 2. Calcareous Spar, or Rhombohedral Limestone. 3. Dolomite, or Macrotypous Limestone. 4. Breunnerite, or Brachytypous Limestone. 5. Ankerite, or Paratomous Limestone.

Order II.—BARYTE.

Genus I. Sparry Iron.—1. Rhombohedral Sparry Iron.

Genus II. RED MANGANESE.—1. Rhombohedral Red Manganese.

Genus III. CALAMINE.—1. Prismatic Calamine, or Silicate of Zinc. 2. Rhombohedral Calamine, or Carbonate of Zinc.

Genus IV. TUNGSTEN, or SCHEELIUM.—1. Pyramidal Tungsten.

Genus V. BARYTE.—1. Strontianite, or Peritomous Baryte. 2. Witherite, or Diprismatic Baryte. 3. Heavy Spar, or Prismatic Baryte. 4. Celestine, or Prismatoidal Baryte.

Genus VI. LEAD-SPAR.—1. White Lead-Spar, or Carbonate of Lead; Diprismatic Lead-Spar. 2. Green and Brown Lead-Spar, or Phosphate of Lead; Rhombohedral Lead-Spar. 3. Red Lead-Spar, or Chromate of Lead; Hemiprismatic Lead-Spar. 4. Yellow Lead-Spar, or Molybdate of Lead; Pyramidal Lead-Spar; 5. Sulphate of Lead, or Prismatic Lead-Spar.

Order III.—KERATE.

Genus I. CORNEOUS SILVER, or CHLORIDE OF SILVER.—1. Hexahedral Corneous Silver.

Genus II. CORNEOUS MERCURY, or DICHLORIDE OF MERCURY.—1. Pyramidal Corneous Mercury.

Order IV.—MALACHITE.

Genus I. COPPER GREEN.—1. Uncleavable Copper Green, or Chrysocolla.

Genus II. LIRICONITE.—1. Prismatic Liriconite, or Lenticular Arseniate of Copper. 2. Hexahedral Liriconite, or Cubical Arseniate of Iron.

Genus III. OLIVENITE.—1. Prismatic Olivenite, or Prismatic Arseniate of Copper. 2. Di-Prismatic Olivenite, or Phosphate of Copper.

Genus IV. BLUE MALACHITE, or BLUE COPPER.—1. Prismatic Blue Malachite. * *Velvet Blue Copper.*

Genus V. EMERALD MALACHITE.—1. Rhombohedral Emerald Malachite, or Dioptase.

Genus VI. GREEN MALACHITE.—1. Prismatic Green Malachite, or Hydrous Phosphate of Copper. 2. Hemiprismatic Green Malachite, or Common Malachite. 3. Prismatoidal Green Malachite, or Muriate of Copper.

Order V.—MICA.

Genus I. Copper Mica.—1. Rhombohedral Copper-Mica, or Micaceous Arseniate of Copper.

Genus II. Uran Mica, or Uranite.—1. Pyramidal Uran-Mica. * Uran-Ochre.

Genus III. Cobalt Mica, or Red Cobalt.—1. Prismatic Red Cobalt. * Cobalt Ochre.—1. Black Cobalt-Ochre. 2. Brown Cobalt-Ochre. 3. Yellow Cobalt-Ochre.

Genus IV. Antimony Mica, or White Antimony.—1. Prismatic White Antimony. * Antimony Ochre.

Genus V. Blue Iron, or Iron-Mica.—1. Prismatic Blue-Iron.

Genus VI. Graphite.—1. Rhombohedral Graphite.

Genus VII. Talc-Mica.—1. Prismatic Talc-Mica, or Talc. 1. *Native Magnesia, or Hydrate of Magnesia ;* 2. *Ophite, or Precious Serpentine ;* 3. *Nephrite ;* 4. *Steatite, or Soapstone ;* 5. *Figure-stone, or Agalmatolite ;* 6. *Magnesite ;* 7. *Meerschaum ;* 8. *Lithomarge ;* 9. *Mountain Soap ;* 10. *Bole.*—2. Rhombohedral Talc-Mica : Mono-axial Mica. 3. Hemiprismatic Talc-Mica, or Common Mica ; Di-axial Mica.

Genus VIII. Pearl-Mica.—1. Rhombohedral Pearl-Mica.

Order VI.—SPAR.

Genus I. Schiller-Spar.—1. Common Schiller-Spar, or Diatomous Schiller-Spar. 2. Bronzite, or Hemiprismatic Schiller-Spar. 3. Hypersthene, or Prismatoidal Schiller-Spar. 4. Anthophyllite, or Prismatic Schiller-Spar.

Genus II. Kyanite.—1. Prismatic Kyanite.

Genus III. Spodumene.—1. Prismatic Spodumene.

Genus IV. Prehnite.—1. Axotomous Prehnite.

Genus V. Datolite.—1. Prismatic Datolite.

Genus VI. Zeolite.—1. Trapezoidal Zeolite, or Leucite. 2. Dodecahedral Zeolite, or Lapis Lazuli. 3. Hexahedral Zeolite, or Analcime. 4. Paratomous Zeolite, or Cross-stone. 5. Rhombohedral Zeolite, or Chabasite. 6. Diatomous Zeolite, or Laumonite. 7. Prismatic Zeolite, or Mesotype. 8. Prismatoidal Zeolite, or Stilbite. 9. Hemiprismatic Zeolite, or Heulandite. 10. Pyramidal Zeolite, or Apophyllite. 11. Macrytopous Zeolite, or Brewsterite.

Genus VII. Petalite.—1. Prismatic Petalite.

Genus VIII. Felspar.—1. Rhombohedral Felspar, or Nepheline. 2. Prismatic Felspar, Common Felspar, or Potash Felspar. 3. Tetarto-Prismatic Felspar, or Soda Felspar. 4. Polychromatic Felspar, or Lime Felspar. 5. Pyramidal Felspar, Scapolite, Meionite.

Genus IX. Augite.—1. Augite; Paratomous Augite. 2. Hornblende; Hemiprismatic Augite. 3. Epidote; Prismatoidal Augite. 4. Wollastonite; Prismatic Augite.

Genus X. Azure-Spar.—1. Prismatic Azure-Spar, or Lazulite. 2. Prismatoidal Azure-Spar, or Blue Spar.

Order VII.—GEM.

Genus I. Andalusite.—1. Prismatic Andalusite.

Genus II. Corundum.—1. Dodecahedral Corundum, or Spinel. 2. Octahedral Corundum, or Automolite. 3. Rhombohedral Corundum, or Sapphire. 4. Prismatic Corundum, or Chrysoberyl.

Genus III. Diamond.—1. Octahedral Diamond.

Genus IV. Topaz.—1. Prismatic Topaz.

Genus V. Emerald.—1. Prismatic Emerald, or Euclase. 2. Rhombohedral Emerald.

Genus VI. Quartz.—1. Prismatic Quartz, or Iolite. 2. Rhombohedral Quartz. 3. Uncleavable Quartz. 4. Fusible Quartz.

Genus VII. Axinite.—1. Prismatic Axinite.

Genus VIII. Chrysolite.—1. Prismatic Chrysolite. 2. Hemiprismatic or Chondrodite.

Genus IX. Boracite.—1. Octahedral Boracite.

Genus X. Tourmaline.—1. Rhombohedral Tourmaline.

Genus XI. Garnet.—1. Pyramidal Garnet, or Vesuvian. 2. Tetrahedral Garnet, or Helvine. 3. Dodecahedral Garnet. 4. Prismatoidal Garnet, or Grenatite.

Genus XII. Zircon.—1. Pyramidal Zircon.

Order VIII.—ORE.

Genus I. Melane Ore.—1. Prismatic Melan-Ore, or Gadolinite. 2. Anorthetic Melan-Ore, or Allanite.

Genus II. Titanium-Ore.—1. Hemiprismatic Titanium-Ore, or Sphene. 2. Octahedral Titanium-Ore, or Pyrochlor. 3. Peritomous Titanium-Ore, or Rutile. 4. Pyramidal Titanium-Ore, or Octahedrite. 5. Prismatic Titanium-Ore, or Brookite.

Genus III. Zinc-Ore.—1. Prismatic Zinc-Ore, or Red Zinc-Ore.

Genus IV. Red Copper-Ore.—1. Octahedral Red Copper-Ore.

Genus V. Tin-Ore.—1. Pyramidal Tin-Ore.

Genus VI. Wolfram-Ore.—1. Prismatic Wolfram-Ore.

Genus VII. Tantalum-Ore.—1. Prismatic Tantalum-Ore.

Genus VIII. Fergusonite.—1. Pyramidal Fergusonite.

Genus IX. Uranium-Ore.—1. Uncleavable Uranium-Ore.

Genus X. Cerium-Ore.—1. Rhombohedral Cerium-Ore.

Genus XI. Chrome-Ore.—1. Octahedral Chrome-Ore, or Chromat of Iron.

Genus XII. Iron-Ore.—1. Octahedral Iron-Ore, or Magnetic Iron-Ore. 2. Rhombohedral Iron-Ore, or Red Iron-Ore. 3. Prismatic Iron-Ore, or Brown Iron-Ore. 4. Diprismatic Iron-Ore, or Lievrite.

Genus XIII. Manganese-Ore.—1. Pyramidal Manganese-Ore, or Black Manganese-Ore. 2. Brachytypous Manganese-Ore, or Braunite. 3. Uncleavable Manganese-Ore, or Psilomelane. 4. Prismatoidal Manganese-Ore, or Grey Manganese-Ore. 5. Prismatic Manganese-Ore, or Pyrolusite.

Order IX.—NATIVE METAL.

Genus I. Arsenic.—1. Rhombohedral Arsenic.

Genus II. Tellurium.—1. Rhombohedral Tellurium.

Genus III. Antimony.—1. Rhombohedral Antimony. 2. Prismatic Antimony, or Antimonial Silver.

Genus IV. Bismuth.—1. Octahedral Bismuth.

Genus V. Mercury.—1. Liquid Native Mercury. 2. Dodecahedral Mercury, or Native Amalgam.

Genus VI. Silver.—1. Hexahedral Silver.

Genus VII. Gold.—1. Hexahedral Gold.

Genus VIII. Iridium.—1. Rhombohedral Iridium.

Genus IX. Palladium.—1. Octahedral Palladium.

Genus X. Platina.—1. Hexahedral Platina.

Genus XI. Iron.—1. Octahedral Iron.

Genus XII. Copper.—1. *Octahedral Copper.*

Order X.—PYRITES.

Genus I. Nickel Pyrites, or Copper-Nickel.—1. Prismatic Nickel Pyrites.

Genus II. Arsenic Pyrites.—1. Axotomous Arsenic Pyrites. 2. Prismatic Arsenic Pyrites.

Genus III. Cobalt-Pyrites.—1. Hexahedral Cobalt-Pyrites, or Silver-White Cobalt. 2. Octahedral Cobalt-Pyrites, or Tin-White Cobalt. 3. Isometric Cobalt-Pyrites. 4. Eutomous Cobalt-Pyrites, or Nickeliferous Cobalt-Pyrites.

Genus IV. Iron Pyrites.—1. Hexahedral Iron-Pyrites. 2. Prismatic Iron-Pyrites. 3. Rhombohedral Iron-Pyrites, or Magnetic Pyrites.

Genus V. Copper-Pyrites.—1. Pyramidal Copper-Pyrites, or Yellow Copper-Pyrites. 2. Octahedral Copper-Pyrites, or Variegated Copper.

Order XI.—GLANCE.

Genus I. Copper-Glance.—1. Tetrahedral Copper-Glance, or Grey and Black Copper. 2. Prismatoidal Copper-Glance. 3. Prismatic Copper-Glance, or Vitreous Copper. 4. Diprismatic Copper-Glance, or Bournonite.

Genus II. Silver-Glance, or Vitreous Silver.—1. Hexahedral Silver-Glance.

Genus III. Galena, or Lead-Glance.—1. Hexahedral Galena, or Lead-Glance.

Genus IV. Tellurium-Glance, or Black Tellurium.—1. Pyramidal Tellurium-Glance.

Genus V. Molybdena, or Molybdena Glance.—1. Rhombohedral Molybdena-Glance.

Genus VI. Bismuth-Glance.—1. Prismatic Bismuth-Glance.
Bismuth-Ochre.

Genus VII. Antimony-Glance.—1. Prismatic Antimony-Glance; Schrifterz. 2. Prismatoidal Antimony-Glance, or Grey Antimony. 3. Axotomous Antimony-Glance; Jamesonite. 4. Peritomous Antimony-Glance, or Sulphuret of Silver and Antimony.

Genus VIII. Melane-Glance.—1. Prismatic Melane-Glance, Brittle Silver-Glance of Werner.

Order XII.—BLENDE.

Genus I. Manganese-Blende.—1. Hexahedral Manganese-Blende.

Genus II. Zinc-Blende, or Garnet-Blende.—1. Dodecahedral Zinc-Blende.

Genus III. Antimony-Blende, or Purple Antimony.—1. Prismatic Antimony-Blende, or Purple Antimony.

Genus IV. Ruby-Blende.—1. Rhombohedral Ruby-Blende, or Red Silver. 2. Hemiprismatic Ruby-Blende, or Cinnabar.

Order XIII.—SULPHUR.

Genus I. Sulphur.—1. Prismatoidal Sulphur, or Yellow Orpiment. 2. Hemiprismatic Sulphur, or Red Orpiment. 3. Prismatic Sulphur, or Common Sulphur.

CLASS III.

Order I.—RESIN.

Genus I. Mellilite, or Honey-Stone.—1. Pyramidal Mellilite, or Honey-Stone.

Genus II. Mineral Resin.—1. Yellow Mineral Resin, or Amber. 2. Black Mineral Resin.

Order II.—COAL.

Genus I. Mineral Coal.—1. Bituminous Mineral-Coal. 2. Glance-Coal.

MINERALOGY.

MINERALOGY is that department of Natural History which makes us acquainted with all the properties and relations of minerals. It is divided into two grand branches, viz. Mineralogy properly so called, and Geology. Mineralogy treats of the properties and relations of *Simple Minerals ;* while Geology considers the various properties and relations of the *Atmosphere*, the *Waters of the Globe*, the *Mountain Rocks*, or those mineral masses of which the crust of the earth is principally composed, and the *form*, *density*, *heat*, *electricity*, and *magnetism of the earth.*

1. *System of Arrangement.*

The arrangement of simple minerals has always been a subject of controversy with mineralogists; and the discussions to which it has given rise have materially contributed to the advancement of our knowledge of the natural and chemical history of minerals. Berzelius contends for the Chemical arrangement, according to which the species are grouped in conformity with their chemical composition and characters; Werner rejects the pure chemical, and adopts the Mixed method, in which the species are arranged and determined according to the conjoined external and chemical characters; of late years the plan, or the arrangement according to external characters alone, (named the Natural History System), has been advocated by Mohs, and also in our System of Mineralogy (in three volumes), and Manual of Mineralogy; and, in our lectures in the University of Edinburgh, we have, for some years, taught our pupils according to a method which, did the limited nature of this article permit, might be more particularly noticed.[1] The Natural History system having been formerly adopted in this work, as one founded on those principles that direct the botanist and zoologist in their determination of the species of plants and animals, and therefore superior in every respect to the plans of Leonhard, Phillips, and others, in which there are no efficient means provided for enabling the student of natural history to refer the minerals he is examining to their place in the system, is still adhered to.

2. *Characters of Minerals.*

As the characters employed in the natural history system of mineralogy are principally those named external, our attention will here be chiefly directed to the consideration of their various properties, using the chemical characters only in so far as they illustrate the constituent parts of the mineral species. We shall first give an account of the characters used in the construction of the different principal divisions of the system; and next enumerate and define those characters which are employed in the *description* of the species, sub-species, kinds, and varieties of simple minerals.

[1] We have in the press a work on Mineralogy and Petralogy, in which the arrangements of the System are founded on External and Chemical characters; written chiefly with the view of meeting the wishes of the medical student and the practical mineralogist and geologist.

I.

CHARACTERS EMPLOYED IN THE DETERMINATION OF THE CLASSES, ORDERS, GENERA, AND SPECIES.

1. Regular Form.—2. Cleavage.—3. Hardness.—4. Specific Gravity.[1]

I. Regular Form.

1. *Simple and Compound Forms.*

The first step Crystallography takes in the consideration of the regular forms occurring in the mineral kingdom, is to distinguish them into *simple* and *compound* ones. A simple form is contained under *homologous* planes, or such as are equal, similar, and equally disposed in respect to each other; a compound one is contained under planes *not homologous*, or such as are not equal, similar, and equally disposed to each other.

2. *Combinations.*

A compound form is termed a *combination*. The combinations consist of the simple forms, and each face occurring in them, belongs to a particular simple form, in as far as it is not homologous with others.

3. *Edges of Combination.*

The edges, in which the faces of different forms meet or intersect each other, are called *edges of combination*. The

[1] Certain varieties of *colour*, the *lustre*, and also the *streak*, are occasionally used in forming the characters of the systematic divisions.

denomination of the edges between the homologous faces depends upon the simple forms themselves, or upon their peculiar situation, and their other qualities.

4. *Development of the Combinations.*

An object of the greatest importance for crystallography is the illustration of the *development* of the combinations. A combination is developed by shewing of what simple forms it consists, and in what relations those simple forms are to each other. The manner in which this is done is as follows: *First,* The connection existing among certain simple forms is exhibited; and these simple forms are arranged accordingly into series, the members of which, between certain limits, follow constant laws, and are known in all their geometrical relations, if it is possible to determine their distance from a known member, or to assign them their relative position in a series, which contains the same member. The position of a member in the series is determined by considering the situation of the edges of combination; and the development of the latter will, therefore, depend upon two points; first, to make out the kind of the simple forms they contain; and, then, from the observed situation of the edges of combination, to determine to which member of the afore-mentioned series they belong, or which places they occupy in the series.

5. *Preparation for the Development of the Combinations.*

Let fig. 1. be a combination of two rhombohedrons. The faces ADEED, ADE′E′D,....belong to one, the faces DEE″E″E′,...to the other. Let the edges of combination DE,DE′,...be parallel, and the faces of the two forms that meet in them equally inclined to each other.

The plane angles at the apex, or at the solid angle formed by three equal plane angles of the former, are smaller than

Fig. 1.

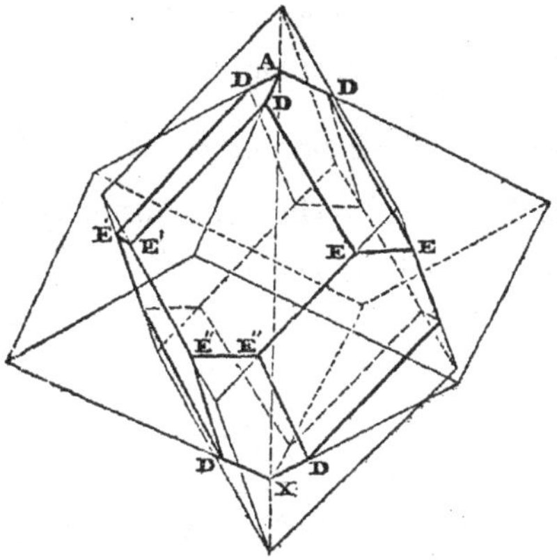

the same of the latter*; and likewise the inclination of the *terminal edges*, or at those which terminate in the apex, or in the terminal point of the axis AX of the former, is smaller than the same of the latter, which, therefore, is called the *more obtuse*, whilst the former bears the denomination of the *more acute rhombohedron*. It would not be sufficiently intelligible to derive the relation which exists between the two rhombohedrons, from the figure in its present state, the absolute dimensions remaining unaltered. Let, therefore, the axis of the more obtuse rhombohedrón increase, till it equals that of the more acute one, without any alteration in the angles, and consequently in the parallelism of the edges of combination depending upon them. Whilst the axis increases, that part of the faces of the more obtuse rhombohedron, which is visible in the combination, will become narrower and narrower, till it disappears, when the length of the axis is equal in both the forms. The planes of the more obtuse rhombohedron are in this case tangent to the terminal edges of the more acute one; and the absolute dimensions are now in that state, in which they allow of a clear development of the relations between the forms which the combination contains.

6. *Upright Position and Horizontal Projection.*

Place a rhombohedron in such a position that its axis is vertical. This is called the *upright position.* From the six angles of the rhombohedron, B′,C′..., fig. 2, which are no apices,

Fig. 2.

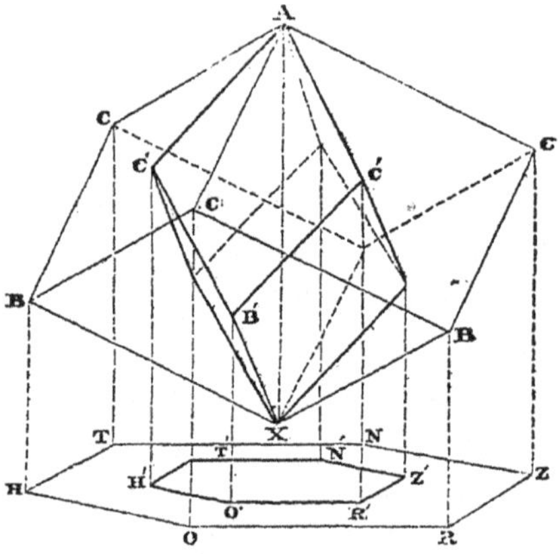

draw perpendicular lines B′O′,C′R′...to a horizontal plane, and join the points in which these perpendiculars meet with the plane, by straight lines O′R′,R′Z′...The regular hexagon H′O′R′Z′N′T′ produced by those lines is called the *horizontal projection* of the rhombohedron.

7. *Relation of the Two Rhombohedrons expressed by the Side of the Horizontal Projection.*

By comparing the planes of the horizontal projections HORZNT and H′O′R′Z′N′T′ belonging to the rhombohedrons we find that, the axes being supposed equal, the horizontal projection of the more obtuse is four times as large as that of the more acute; or it is to the same in the ratio of 4 : 1. From this follows the ratio of their sides = 2 : 1. Therefore, two rhombohedrons, the horizontal projections of which are in the ratio of 4 : 1, or the sides of these in that of 2 : 1, the axes being supposed equal, will be able to produce a combination in which the edges of combination are parallel to each other; also to the terminal edges of the more acute, and to the inclined diagonals (those which join the apex with the opposite angle of the faces) of the more obtuse rhombohedrons; provided the absolute dimensions allow the space to be limited by both the forms at the same time. This is one of the relations between simple forms, which are useful and necessary for explaining the combinations, and of which the other relation will be explained in the following paragraphs.

8. *Fundamental Form, Derivation, Derived Form.*

What has been said above leads to a general method, one form being given or chosen, to find one or several others, which are in such relations to it, as to be fit for explaining the combinations. The form chosen for this purpose is called the *fundamental form* or *type;* the proceeding by which the other forms are produced the *derivation*, and those forms themselves the *derived forms.*

9. *Relation of the Two Rhombohedrons expressed by the Axis.*

The easiest method of expressing the relation of the derived rhombohedrons to the fundamental form (§ 7.) is not that by the sides of the horizontal projections, the axes being supposed equal. Let, therefore, the derived rhombohedron, without any alteration in its angles, diminish in size (as could be done by sections parallel to its faces) till its horizontal projection becomes equal to the horizontal projection of the fundamental form; its axis will decrease to half of its former length. Now, the horizontal projections being supposed equal, the axis of the derived rhombohedron will be to the axis of the fundamental form in the ratio of $\frac{1}{2} : 1 = 1 : 2$, an expression of the relation between the two forms, which, on account of its easy application, we shall retain and make use of as we proceed.

10. *Inversion of the Proceeding.*

The mode of proceeding, by which the more obtuse rhombohedron has been derived from the more acute one, can easily be inverted; setting out from the more obtuse one, we can, by the same way, in an opposite direction, derive the more acute rhombohedron. The diagonals being drawn upon the faces of the rhombohedron, and by planes laid through three and three of them, those parts which are situated on their outside being taken away, what remains will be a rhombohedron more acute than the fundamental form. The axes of both these forms are equal, whilst the sides of the horizontal projections in the fundamental and the derived rhombohedron are in the ratio of $1 : \frac{1}{2}$. Now, the horizontal projections being equal, the ratio of the axes will be $= 1 : 2$.

11. *Generality of the Relations.—Series.*

These modes of proceeding, and the relations between the derived forms, are general. If, therefore, from one rhombohedron, taken for the fundamental form, another more obtuse has been derived; this same more obtuse rhombohedron, considered as a fundamental form, will yield another still more obtuse; and if, on the other hand, from the fundamental form a more acute rhombohedron has been derived, the same proceeding applied to the resultant form will give another still more acute; and this may be continued on either side *ad infinitum*. The ratio between the axes of any two subsequent rhombohedrons, the horizontal projections being supposed equal, has every where the same exponent. The axes are therefore in a geometrical progression, and the rhombohedrons represented by these axes are themselves said to form such a progression, or a *series*.

12. *Laws of Progression of the Series.*

Let A be the fundamental form, its axis $= a$. Let B represent the first, C the second more obtuse; B′ the first, C′ the second more acute rhombohedron, the rhombohedrons will follow each other thus;

$$\ldots C, B, A, B', C' \ldots$$

their axes will be

$$\ldots \frac{a}{4}, \frac{a}{2}, a, 2a, 4a \ldots$$

or they are in the ratios of

$$\ldots \tfrac{1}{4}, \tfrac{1}{2}, 1, 2, 4 \ldots$$

$$\ldots 2^{-2}, 2^{-1}, 2^0, 2^1, 2^2 \ldots$$

equal to those of the powers of the number 2.

The more obtuse rhombohedrons B, C ... and the more acute ones, B′, C′ ... derived from a given or supposed fundamental form A, their horizontal projections being equal, produce a series, in which the axes of the subsequent members increase on one side, and decrease on the other, like the powers of the number 2.

13. *Designation.*

The method of designating crystallizations, whether simple or compound, is founded on the present and similar series, to be developed afterwards. Represent the fundamental form by any arbitrary letter, in this case, according to the practice hitherto observed, by R. The sign of any other member of the series contains the same letter; the exponent of the fundamental number of the series, which, in the present case, is the number 2, being added for the sake of expressing the place occupied in the series by the number to be designated. This yields the crystallographic sign for one member, and by applying the same proceeding to the others, the designation also of the whole series.

14. *Continuation.*

The above mentioned part of the series of rhombohedrons is

C, B, A, B′, C′;

the ratio of their axes

$$2^{-2},\ 2^{-1},\ 2^{0},\ 2^{1},\ 2^{2};$$

its designation, therefore,

$$R-2,\ R-1,\ R,\ R+1,\ R+2,$$

and in the same way $R+n$, the general term for each member of the series, which may be transformed into a particular one, by putting whole numbers, positive or negative, instead of n in the expression.

15. *Usefulness of the Designation.*

From the ratio of the axis to the side of the horizontal projection being known in a rhombohedron, all the rest of its ratios, its dimensions (plain and solid angled), &c. can be deduced. The side of the horizontal projection is supposed equal in all the members of the series, and made = 1 in the calculations which refer to the rhombohedron. The designation contains therefore, by representing the relation of the axes, every thing required as a basis of calculation; and since it does not require any figures (for instance those of a fundamental form), the letter R signifying by itself a rhombohedron, the designation is also evident, and upon this depends the easiness of its application in crystallography.

16. *Limits of the Series.*

There can be no doubt of n being able to obtain any possible value; or, which is the same, the series may be continued as far as we choose. This produces always new rhombohedrons, as long as n has a finite value. On one side the axis will increase, if n increase and remain positive; on the other the axes will decrease, if the negative value of n increases. Let n become infinite, the axis will likewise be infinite; let $-n$ become infinite, the axis will be infinitely small. It is clear that the limits of the series will be, on one side, a rhombohedron of an infinitely large axis, on the other, a rhombohedron of an infinitely small axis, and their several signs must therefore be $R+\infty$ and $R-\infty$; so that

$$R-\infty \ \ldots \ R+n \ \ldots \ R+\infty$$

represent the series between its limits.

17. *The Limits are Regular Six-sided Prisms.*

It is very easily shewn that a rhombohedron of an infinite axis is a regular six-sided prism; for at the same time, as the axis increases, the six faces of the rhombohedron approach nearer and nearer to parallelism with the axis, and they become really parallel with it, if the axis becomes infinitely long. For rendering more evident in what manner the regular six-sided prism is the limit of the series of rhombohedrons, we can imagine each face of the rhombohedron to make a revolution round an immoveable line QS, fig. 3. This

Fig. 3.

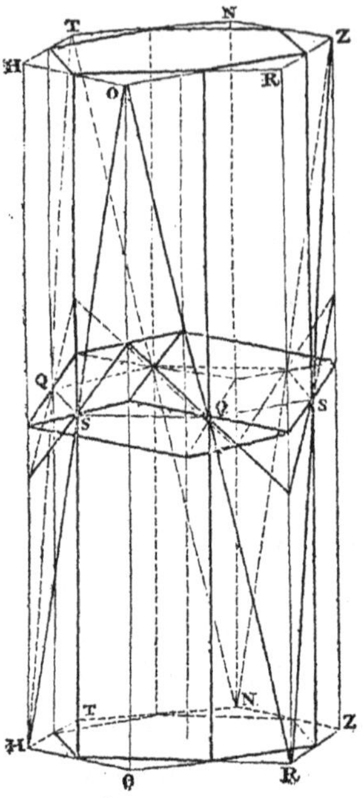

line is the side of that regular hexagon, which can be inscribed into the horizontal projection HORZNT; the hexagon itself will be the basis or transverse section of the prism, and it differs from the horizontal projection by both its contents and its situation. If the axis of the rhombohedron decrease till it ultimately evanesce, the faces contiguous not only to one, but also those contiguous to the other apex, coincide in one plane; and the rhombohedron is changed into a plane figure, equal and similar to the horizontal projection, to which it is parallel, and appears, therefore, in all the combinations as a face perpendicular to the axis. It must here be remarked, that forms of infinite dimensions cannot be produced by themselves, or as simple forms, but appear only in combinations.

18. *Position.*

The series of rhombohedrons, already described, contains every thing concerning the dimensions and the ratios of its members, as far as each of them is considered singly or by itself. But there is still one object of great importance to be noticed, if those members are considered connected together, or in combinations. This is their *position*. It is evident that two subsequent members of the series, as, for instance, R and R+1, or, in general terms, $R+n$ and $R+n+1$, will produce the above combination in which the edges of combination are parallel, not only with each other, but also with the terminal edges of the more acute and with the inclined diagonals of the more obtuse rhombohedrons; but they will produce this combination only in the case of their being in such a position, that the terminal edges of the one are in the same plane with the inclined diagonals of the other, which plane passes at the same time through the axis, as is shewn by fig. 2. This position of two rhombohedrons is termed the *turned position*, because it arises when a rhombohedron is turned out of its former position, under an angle of 60° or 180°. If a rhombohedron, in the turned position, undergo a new revolution of 60° or 180°, it returns to that which it occupied before, and this is called the *parallel position*. Rhombohedrons that affect a parallel position with each other, have their terminal edges and their inclined diagonals distributed in such a manner, that the terminal edges of both on one side, and the inclined diagonals of both on the other, fall into one and the same plane, passing through the axis.

19. *Position of the Members in respect to each other.*

The position of the fundamental form R is considered as the *normal position*, to which those of the derived members are referred. In reference to R, R+1 and R—1 will be in a turned, R+2 and R—2 in a parallel, R+3 and R—3 in a turned position again, and so on; from which it follows, that any two subsequent, or, in general, any two members of the series, between which an even number of members (0 in the above case being considered as even) are wanting, will be in the turned position, whilst any two members of the series, between which an odd number of members are wanting, will affect the parallel position. All those members, therefore, which have their exponent an even number, are in the parallel, those which have their exponent an odd number in the turned position, if compared with R, the fundamental form of the series. The position exercises no influence upon the regular six-sided prism, since, by turning it under angles of 60° and 180°, its faces will come into their former situation.

20. *Scalene Six-sided Pyramids.*

The rhombohedron allows several forms to be derived from it, which are not themselves rhombohedrons like the original. Lengthen the axis of a rhombohedron, fig. 4, on both sides to an indefinite but equal distance, and from the angles B, C, ... draw straight lines to the terminal points A′, X′, of the lengthened axis. The planes which can be laid through any two contiguous lines of these, and the lateral edges,

BC, CB,...of the rhombohedron will produce a form, which, from its faces being scalene triangles, is called a *scalene*

Fig 4.

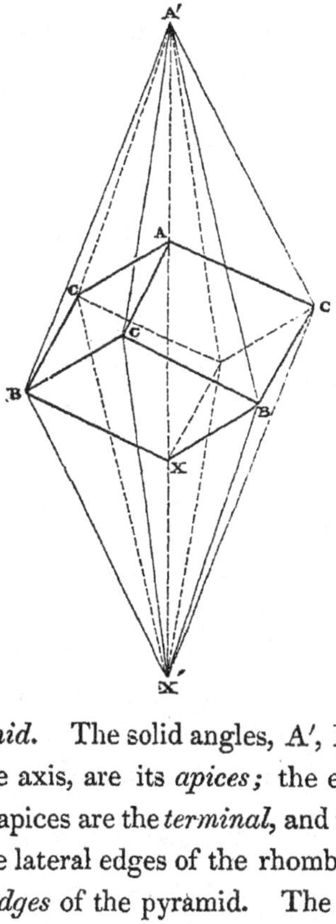

six-sided pyramid. The solid angles, A′, X′, at the terminal points of the axis, are its *apices;* the edges which terminate in these apices are the *terminal*, and those which correspond with the lateral edges of the rhombohedron are the *rhombohedral edges* of the pyramid. The terminal edges allow of a farther distinction into *more obtuse* and *more acute* ones ; the faces being on AB, the first, inclined under a greater, on AC, the other, under a less angle.

21. *Their Axis a Multiple of the Axis of the Fundamental Form.*

Instead of lengthening the axis, we may conceive it to

be multiplied by a certain number, expressed in general by *m*. The values of *m* must be determined by observation, or, which is the same, they must be derived from combinations containing six-sided pyramids. They cannot, therefore, be fixed arbitrarily, if we expect them to be confirmed by observation.

22. *Series of Scalene Six-sided Pyramids.*

From each rhombohedron, several scalene six-sided pyramids may be derived, and we are authorized to expect as many forms of this kind as observation has discovered different values for *m*. By deriving scalene six-sided pyramids, according to a constant *m*, from the subsequent members of the series of rhombohedrons, a *series of such pyramids* is produced, the axes of which partake of the law followed by those of the series of rhombohedrons. For the horizontal projections of the pyramids are identical with those of the rhombohedrons, because the rhombohedral edges of the pyramids have the same situation as the lateral edges of the rhombohedrons from which they are derived; and the axes of the pyramids, being equal to the axes of the rhombohedrons, multiplied by a constant number *m*, will therefore persist in the ratio of those.

23. *Designation.*

The letter P, in general, signifies a pyramid, but more particularly a six-sided pyramid, if referred to a form derivable from the rhombohedron. The quality of a scalene six-sided pyramid, in which there occur differences according to the relation of its axis to that of the rhombohedron, must be made evident by the designation. Include the general sign for the pyramid, expressed in such a manner as to refer to the rhombohedron from which it is derived, within crotchets (), and add the number of derivation, or that by which the axis of the rhombohedron has been multiplied, in order to produce the axis of the pyramid, in the form of an exponent; the result will be the crystallographic sign of a scalene six-sided pyramid. Thus $(P+1)^3$ represents a scalene six-sided pyramid derived from R+1 by multiplying its axis by the number 3; and $(P+n)^m$ is the general term of any or of each series of scalene six-sided pyramids, which can be derived from a series of rhombohedrons having R for its fundamental form, according to the value of *m*, determined for each particular series. If the axis of the rhombohedron from which a scalene six-sided pyramid is derived, and the number of derivation *m* are known, the dimensions of the pyramid are easily to be found; thus every thing that has been said above (§ 15.) in respect to the designation, can be equally applied in the present case.

24. *Values of m.*

Although the value of *m* be, as far as experience goes, most commonly = 2, 3, or 5, yet it sometimes affects others, and even fractionary numbers. The only thing to be said in general of the value of *m* is, that it must be greater than 1, rational and positive.

25. *Limits.*

Every series leads to its limits; it is, therefore, to be expected, that the series of scalene six-sided pyramids will have limits of their own. It appears by an easy geometrical construction, that scalene six-sided pyramids, derived according to a constant *m*, from the different members of a series of rhombohedrons, or even from any rhombohedron though it be no member of that series, if cut perpendicularly upon, and through the centre of their axes; that such pyramids will have a common, or equal and similar transverse section, provided only the horizontal projections of these rhombohedrons have been equal. This section is an equilateral dodecagon of alternately equal angles. The limits of a series of scalene six-sided pyramids must therefore be a scalene six-sided pyramid of an infinite axis; which, since m cannot become infinite, must be derivable from a rhombohedron of an infinite axis, that is to say, from a regular six-sided prism. The limit will, therefore, appear in the shape of an *unequiangular twelve-sided* prism, of the same transverse section, which the finite members of the series produce. Each series of scalene six-sided pyramids is limited by a prism of this kind; the number of different prisms will therefore be equal to the values of m, upon which the angles of their transverse sections depend. The opposite limit, viz. the scalene six-sided pyramid of an infinitely small axis again appears in a plane figure, equal and similar to the horizontal projection, its faces being perpendicular to the axis of those forms with which it is combined. The general sign of an unequiangular twelve-sided prism is $(P+\infty)^m$. The face perpendicular to the axis does not require any particular designation, since that of $R-\infty$ (§ 15) is already applied to it. Respecting the position, both of the scalene six-sided pyramids, and of their limits, the unequiangular twelve-sided prisms to each other, and to the rhombohedral form from which they are derived; §§ 18 and 19 contain the necessary explanations.

26. *Subordinate Series of Rhombohedrons.*

There are several rhombohedrons in connection with R,

without being members of the series derived from it, because their axes, the horizontal projections being equal, are not products of the axes of R, multiplied by powers of the number 2. These rhombohedrons follow, however, the law (§ 12) which governs the series of rhombohedrons immediately derived from R, and their axes are results of the axes of the members of this series multiplied by a certain coefficient. Series arising in this way are termed *subordinate series*, whilst that which has been derived from R has the relative name of the *principal series*.

27. *Coefficients of the Subordinate Series, and how the Members of these belong to each other, and to the Members of the Principal Series.*

Members of the subordinate series are geometrically obtained by laying planes into the homologous terminal edges of the scalene six-sided pyramids derived from the members of the principal series, and by enlarging these planes till they enclose the space by themselves. By applying the planes to the more obtuse terminal edges, the resulting coefficient is $= \frac{3m+1}{4}$; by applying them to the more acute ones, it follows $= \frac{3m-1}{4}$. If the three most common values of m are one after the other put into the expressions, each of the two coefficients give rise to members of two different subordinate series; but at the same time it yields one member of the principal series, namely, if $3m \pm 1$ becomes equal to a power of the number 2. These numbers of the principal series are not, however, objects of the present consideration. The subordinate series, whose coefficient contains, besides a power of the number 2, the number 5, is styled the *first;* that whose coefficient, also besides a power of the number 2, contains the number 7, the *second* subordinate series. The members of the subordinate series arrange themselves with those of the principal series in such a manner, that the axes of those which are put together stand in the ratio of $1 : \frac{5}{4} : \frac{7}{4}$. According to this disposition, a rhombohedron, whose axis contains five-fourths of the axis of $R+n$, or which is $= \frac{5}{4} . 2^n a$, its horizontal projection being equal to that of R, is $R+n$ of the first subordinate series, and its designation $\frac{5}{4}R+n$; a rhombohedron whose axis is $= \frac{7}{4} . 2^n a$, is $R+n$ of the second subordinate series, and its designation is $\frac{7}{4}R+n$. Any other value of m affords other subordinate series. The members of a subordinate series cannot be interpolated, or put between those of the principal one, without undoing both these series themselves. The limits of the subordinate series are identical with those of the principal series; the position of their members follows from their derivation.

28. *Isosceles Six-sided Pyramids.*

Apply pairs of planes to the terminal edges AC, of a rhombohedron, fig. 5, in such a manner and inclination, that

Fig. 5.

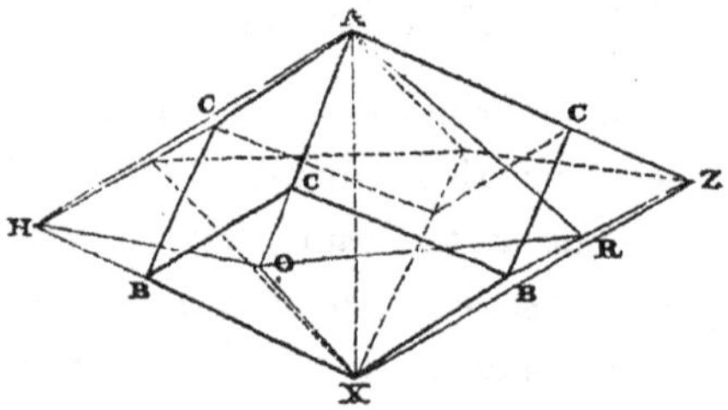

those from the upper apex, AHO and AOR, may intersect those from the under apex, XHO and XOR, in a plane figure HORZNT, similar to and parallel with the horizontal projection. The result will be an *isosceles six-sided pyramid.* The axes of both the pyramid and the rhombohedron being equal, as appears from the derivation, the sides of their horizontal projections are in the ratio of 3 : 2. Hence the horizontal projections being supposed equal, the ratio between the axis of the pyramid and that of the rhombohedron will follow = 2 : 3, or the first will contain $\frac{2}{3}$ of the axis of that rhombohedron from which it is derived. The isosceles six-sided pyramid, derived from R+*n*, is designated only by P+*n*, (§ 23), on account of its being always obtained by a similar proceeding from its rhombohedron.

29. *Series of Isosceles Six-sided Pyramids.*

Each member of the principal series of rhombohedrons has its concomitant isosceles six-sided pyramid of the ratio just stated, the horizontal projections being equal. This gives rise to a series of isosceles six-sided pyramids, the axes of which, their horizontal projections being equal, increase and decrease as the powers of the number 2; that is to say, they follow the general law of all those forms which are derived from the rhombohedron.

30. *Limits and Position.*

The limits of this series are regular six-sided prisms of infinite axes, like those of the series of rhombohedrons. Let the axis of an isosceles six-sided pyramid increase, whilst the horizontal projection remains unchanged, the inclination of the faces at the edges of the basis (those in which the faces from different apices meet) will likewise increase till it becomes = 180°, in which case the axis will be infinite. In this case, each two corresponding faces from the different apices coincide into a single plane parallel with the axis, and the isosceles six-sided pyramid is thus transformed into a regular six-sided prism, the basis or transverse section of which is the horizontal projection itself. By this basis and its situation the present prism differs from that (§ 17), which limits the series of rhombohedrons; and by this the necessity of considering them as two distinct forms becomes evident. The crystallographic sign for the limit of the series of isosceles six-sided pyramids is $P + \infty$. The opposite limit, merely a face perpendicular to the axis of any of the foregoing forms, requires no particular designation. Since the faces of an isosceles six-sided pyramid, when this is turned round under 60° or 180°, resume their former situation, there is no difference prevailing in the position of the members and limits of this series, either in respect to themselves, or to the other forms derived from the rhombohedron.

31. *Derivation from the Isosceles Four-sided Pyramid.*

Consider an *isosceles four-sided pyramid* (fig. 6), a

Fig. 6.

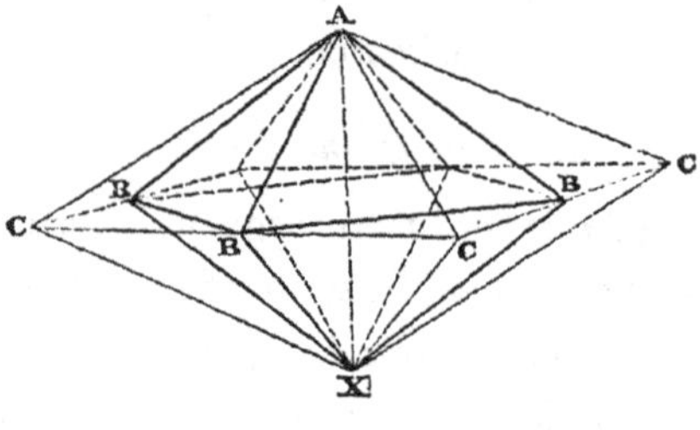

form contained under eight equal and similar isosceles triangles, as a fundamental form; place it in its upright position, by making its axis AX vertical, and apply to it the process § 5. The tangent planes will limit a space representing again an isosceles four-sided pyramid, the plane angles of which at the apex, and the inclination of the faces at the terminal edges, will be greater than the same in the fundamental form, and hence it will likewise be the more obtuse of both.

32. *Ratio of the Derived Pyramid to the Fundamental Form.*

The axes of these pyramids are equal to each other. The horizontal projection of the more obtuse is that square which can be circumscribed to the horizontal projection of the fundamental form; and, accordingly, its content twice the content of the horizontal projection of the fundamental form, the ratio of their sides $= \sqrt{2} : 1$. Suppose now the horizontal projections of the two pyramids to be equal, the axis of the derived more obtuse pyramid will be to the axis of the fundamental form in the ratio of $\frac{1}{\sqrt{2}} = \sqrt{2} : 1$.

33. *Inversion of the Proceeding.*

This proceeding can be applied inversely, in order to obtain from the fundamental form the same more acute four-sided pyramid, from which, by the proceeding applied directly, the fundamental form would have been the result. This inverse application is as follows: Draw in the faces of the fundamental form perpendicular lines from the apices to the edges at the basis, lay cutting planes through every two contiguous of these lines, and separate those parts of the pyramids which lie outside of them. The remainder, inclosed by the cutting planes, will be again an isosceles four-sided pyramid, having the same axis as the fundamental form, but a horizontal projection, whose content is half the content of the horizontal projection of the fundamental form, and the ratio of their sides, therefore, $= \sqrt{\frac{1}{2}} : 1$. The horizontal projections being made equal, the axis of the derived more acute pyramid is to the axis of the fundamental form in the ratio of $\sqrt{2} : 1$, which is the inverse of the ratio deduced above.

34. *Series and Designation.*

By continuing the derivation, the more obtuse pyramid produces another still more obtuse; the more acute another still more acute, and thus arises a series of isosceles four-sided pyramids, whose axes, their horizontal projections being equal, increase and decrease as the powers of the square root of 2; which law of progression can be expressed by $2^{\frac{n}{2}}$ or $\sqrt{2}^n$. The designation of this series is concordant with the principles already laid down and followed (§ 13). The fundamental form receives the letter P Hence its neighbouring members will be

$$\ldots \quad P-2, P-1, P, P+1, P+2, \quad \ldots$$

their axes that of P being $= a$,

$$\ldots \frac{a}{2}, \quad \frac{a}{\sqrt{a}}, \quad a, \quad a\sqrt{a}, \quad 2a; \ldots$$

their ratio to each other,

$$\ldots \quad \sqrt{2}^{-2} + : \sqrt{2}^{1} : \sqrt{2}^{0} : \sqrt{2}^{1} : \sqrt{2}^{2}; \quad \ldots$$

and $P+n$ the general term of the series, which, by the substitution of any whole positive or negative number in the place of n, gives a certain determined member. The present designation is quite the same as that in § 28, and as another to be mentioned in § 44. However, the observation in the section on the "mode of determining a mineral, or referring it to its place in the system," is sufficient to shew the impossibility of any confusion or ambiguity that could arise from making use of identical signs for the expression of different forms.

35. *Limits.*

If *n* becomes greater or smaller than any given number, a transformation takes place similar to that by which the isosceles six-sided pyramid gave rise to the regular six-sided prism: the series of isosceles four-sided pyramids terminates on one side in an unlimited four-sided prism, on the other in a plane figure, equal, similar, and parallel to the horizontal projection, and these two forms represent the limits of the series. In respect to the first, however, there occurs the particular circumstance, that the series is doubly limited; because the different members in their succession assume a different position. The derivation shews the pyramids, P—1 and P+1, to stand in such a position towards P, that the sides of their bases are parallel with the diagonals of the basis of P; whilst in P—2 and P+2, the sides and the diagonals are parallel with the same lines in the basis of the fundamental form. The former are said to be in a *diagonal*, the latter in a *parallel* position with P; and every thing contained in § 18 and § 19, in respect to position, may be applied here, if only, instead of 60° and 180°, a revolution of 45° is imagined. Suppose now the two different prisms, the regular six-sided, as well as the rectangular four-sided, in both their peculiar positions. The faces of the first will, after having been duly turned, assume their former situation, which is not the case in the second; and hence it becomes necessary to adopt two rectangular four-sided prisms of an infinite axis as limits of the series of isosceles four-sided pyramids, one being in a parallel, the other in a diagonal position with P. The sign of the first is $P+\infty$, that of the second $[P+\infty]$. The opposite limit, where there is no difference in the position, is represented by $P-\infty$; and the whole series of isosceles four-sided pyramids between its limits appears thus,

$$P-8 \ldots\ldots P+n \ldots\ldots \begin{Bmatrix} P+\infty \\ [P+\infty] \end{Bmatrix}$$

36. *Scalene Eight-sided Pyramids.*

With the isosceles four-sided pyramids there are connected *scalene eight-sided pyramids*, contained under sixteen equal and similar scalene triangles (fig. 7). These eight-sided

Fig. 7.

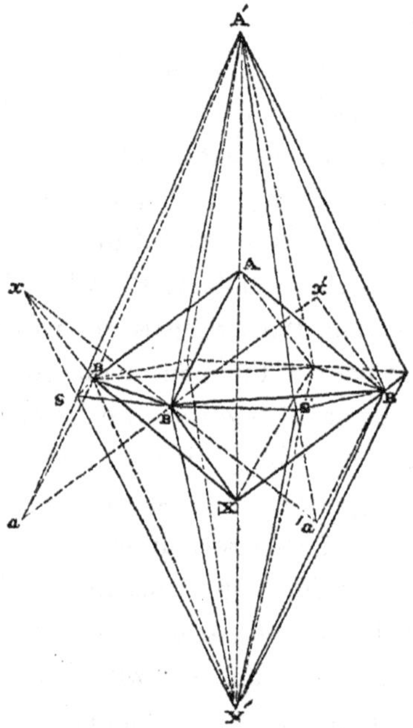

pyramids depend upon the four-sided by the same process of derivation, by which the scalene six-sided pyramids take their rise from the rhombohedron. The application of this process to the isosceles four-sided pyramid supposes, however, a preparation of this form, which is effected by enlarging its faces beyond the edges of its basis, and by drawing in these enlargements triangles, equal and similar to the faces of the pyramid. Thus the points a, a', . . x, x . . are fixed in two squares that are perpendicular to the axis AX, in its terminal points, or in the apices of the pyramids. Now, produce the axis of the pyramid on both sides to an indefinite but equal length, or multiply it by a number m, greater than 1 positive and rational (§ 24.) Join, then, the upper terminal point A′ of the lengthened axis with the points a, a' . . in the under, the under terminal point X′ with the points x, x' . . in the upper square, by straight lines. The points SS′ . ., in which these lines intersect each other, will lie in the prolongation of the basis of the four-sided pyramid. We now join these points with the neighbouring angles of the base by straight lines, the result of which will represent the basis of the derived scalene eight-sided pyramid, which may now be completed without difficulty. The most common, though not the only values given for m by nature, are 3, 4, and 5.

37. *Series of Scalene Eight-sided Pyramids.*

If the number of derivation m be constant, the resultant scalene eight-sided pyramids have similar bases, the axes of the isosceles four-sided pyramids may be whatever they will. The application of the described proceeding to the subsequent members of the series of isosceles four sided pyramids will, therefore, m remaining for all the members the same, not only produce a series of scalene eight-sided pyramids, subject to the law followed by the isosceles four-sided pyramids, which is that of $\sqrt{2}$; but the bases or transverse sections of all the members of this series will be similar to each other. The designation of an indeterminate nth member of such a series is $(P+n)^m$, and it is evident, that as many values of m as experience gives, so many different series of scalene eight-sided pyramids will arise.

38. *Limits.*

The limits of those series are, on one side unlimited unequiangular eight-sided prisms, the transverse sections of them being similar to the bases of the members of the series; on the other side, a plane figure, which in the combinations appears as a face perpendicular to the axis. This is evident from what has been said above, and there remains only to be added, that the prisms of an infinite axis coming into consideration in two different positions, they must be taken for two different forms, or as double limits of the series. These two positions are the *parallel* and the *diagonal,* as in § 35; the parallel being that which prevails between scalene eight-sided pyramids and that isosceles four-sided from which they are derived; whilst forms of this kind in the diagonal position are supposed to have undergone a horizontal revolution of 45°. Now, the eight-sided prisms may be considered as scalene eight-sided pyramids, derived from isosceles four-sided pyramids of an infinite axis, namely, the rectangular four-sided prism, according to a certain value of m. The necessity of considering the four-sided prism in m two different positions (§ 35) must therefore be extended to the eight-sided. The designation of the series of eight-sided pyramids, between their limits, will accordingly be thus:

$$P-\infty \quad \ldots \quad (P+n)^m \quad \ldots \quad \left\{ \begin{matrix} (P+\infty)^m \\ [(P+\infty)^m] \end{matrix} \right\}$$

39. *Subordinate Series.*

There are series of isosceles four-sided pyramids belonging to that of § 33, which, in reference to the latter, or the *principal series*, are termed *subordinate series*. The members of these series are obtained by placing tangent planes in the homologous terminal edges of the scalene eight-sided pyramids; the same proceeding which applied to the scalene six-sided pyramids produced the subordinate series of rhombohedrons. The coefficient for the more obtuse terminal edge (if $m \gtrsim 1+\sqrt{2}$) is $= \frac{m+1}{2}$; that for the more acute one (under the same supposition) $= \frac{m}{\sqrt{2}}$. If m be determined, each of those coefficients gives a member of a particular series. The members of the principal and of the two subordinate series arising from $m = 3$, 4, and 5, arrange themselves thus: $1 : \frac{3}{2\sqrt{2}} : \frac{5}{4}$; and are said to belong together in this order. An isosceles four-sided pyramid, therefore, whose axis, the horizontal projection being equal to that of P, is $= \frac{3}{2\sqrt{2}}$. $\sqrt{2}^n a$ represents the member $P+n$ of the first subordinate series, and bears the designation of $\frac{3}{2\sqrt{2}} P+n$; and a similar pyramid, the other circumstances remaining the same, with an axis $= \frac{5}{4}$. $\sqrt{2}^n a$

is the member $P+n$ of the second subordinate series, expressed by the designation $\frac{5}{4}P+n$. The limits of the present series coincide with those of the principal series. The positions of its members in respect to each other, and to the members of the principal series, will become evident by comparing what has been stated of their derivation with § 27, regarding the position of the subordinate series of rhombohedrons.

40. *Scalene Four-sided Pyramids.*

If in the basis of an isosceles four-sided pyramid, the diagonals are supposed unequal, the pyramid itself is transformed into a *scalene four-sided pyramid*, or into one whose basis is a *rhomb*, its faces *scalene* triangles, and its terminal edges of a different magnitude, the one more obtuse, the other more acute, as is represented by ABCB′C′X, fig. 8.

Fig. 8.

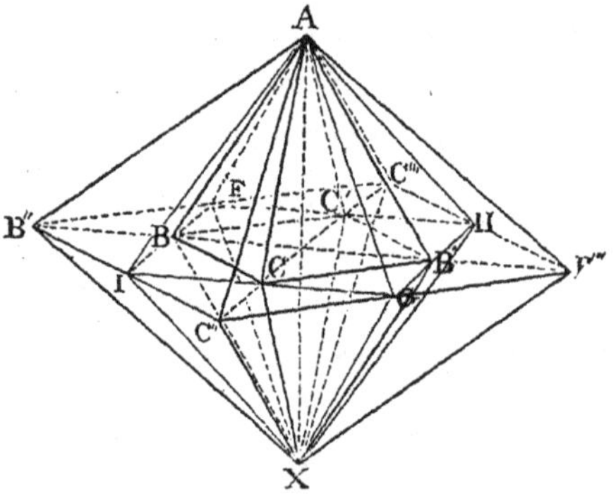

From this we may conjecture that several of the preceding methods of derivation, though with some modifications, will also in this case be applicable.

41. *Derivations from it. Auxiliary Form.*

Apply to the *homologous* terminal edges of a scalene four-sided pyramid tangent planes, and enlarge them till they intersect each other in every possible direction; the result will be no form limited on every side, or such as is finite in all its dimensions. If the tangent planes are laid in *all* the terminal edges of the fundamental form, a form limited from every side will indeed appear, but this form is *not* a *simple* one, its faces being homologous only by four and four (§ 1). The form thus arising represents a four-sided pyramid, with an *oblong rectangular* basis, AIGHFX, fig. 8; and most of the crystallographers consider it as such, and term it accordingly. In this place it must be considered only as an *auxiliary* or *intermediate* form, which is not the derived form itself, but useful and auxiliary for its derivation.

42. *Derived Form.*

Suppose tangent planes to be laid in the terminal edges of the *auxiliary form*, in such an inclination that those from the upper apex produce by their intersection with those from the under apex, a rhomb similar and parallel to the basis of the fundamental form, and situated in the prolongation of this basis. The result will be a scalene four-sided pyramid, the basis of which is similar to that of the fundamental form; but the plane angles at the apex, and the inclination at the terminal edges, being greater than in this form, the derived will be more obtuse than the fundamental form.

43. *Ratio to the Fundamental Form.*

The two pyramids, as produced by the derivation, have

the same axis, the ratio betweeen the basis of the derived, and the basis of the fundamental pyramid being that of 4 : 1. This ratio is evident, since the basis of the auxiliary form is double of the basis of the fundamental form, and that of the derived form again double of the basis of the auxiliary form. Hence the horizontal projection of the two forms being supposed equal, the ratio between the axes of the derived, and of the fundamental pyramid, is that of $\frac{1}{2} : 1$. The process of derivation, by which the more obtuse pyramid is produced, is liable to an inversion like that in § 10.

44. *Series. Designation. Limits. Position.*

From a continued derivation on both sides of the fundamental form, a series of scalene four-sided pyramids, of equal and similar bases, will evidently arise, the axes of which increase on one side and decrease on the other, like the powers of the number 2. $P+n$ will be the sign of an indeterminate nth member of the series, and if n becomes infinite, the series reaches its limits, which are on one side an obliquangular four-sided prism, whose transverse section on the other side is a horizontal plane, whose figure is equal and similar to the horizontal projection of the fundamental form. The designation of the series between its limits becomes thus:

$$P-\infty \ldots P+n \ldots P+\infty .$$

It must be remarked, that such differences on account of the position of the members of the series as are met with in the rhomboid (§ 18) and the isosceles four-sided pyramid (§ 35), are not to be found in the scalene four-sided pyramid.

45. *Further Derivations.*

The members of the present series are, or this series itself is, now the foundation of several other derivations; and although the forms derived by them will be nothing but scalene four-sided pyramids and obliquangular four-sided prisms, yet the variety in the relations of those forms is great enough to surpass that of the forms in connection with the rhombohedron, and with the isosceles four-sided pyramid, for which reason it is one of the most interesting objects of crystallography. The derivations already mentioned apply as well to the fundamental form itself as to the auxiliary form; and they are similar to that which has been employed in deriving the scalene eight-sided from the isosceles four-sided pyramid.

46. *From the Fundamental Form.*

Instead of the isosceles (§ 37), suppose the scalene four-sided pyramid to be the fundamental form, and prepare it for derivation by enlarging its planes beyond the edges at the basis, and by drawing triangles equal and similar to the faces of the pyramid in these enlargements. The points a, a' . . x x' . . will no longer be the angles of squares, but of oblong rectangles, the planes of which are perpendicular to the axis AX in its terminal points. Produce now the axis of the scalene four-sided pyramid on both sides to an indefinite but equal length, or, what comes to the same, multiply it by a number m, enjoying the general properties mentioned, §§ 24, 26. Draw straight lines from the points a, a' . . in the lower rectangle to the upper apex A′, from the points x, x' . . in the upper rectangle to the under apex X′ of the lengthened axis, and join their intersections s, s' . . in the plane of the enlarged basis by straight lines with the angles of this basis. The resultant octagon will not have alternately equal angles, like that in § 36, which is the basis of the scalene eight-sided pyramids, but it will be irregular, fig. 9, since of the angles B, B′, c, c', only the opposite are

Fig. 9.

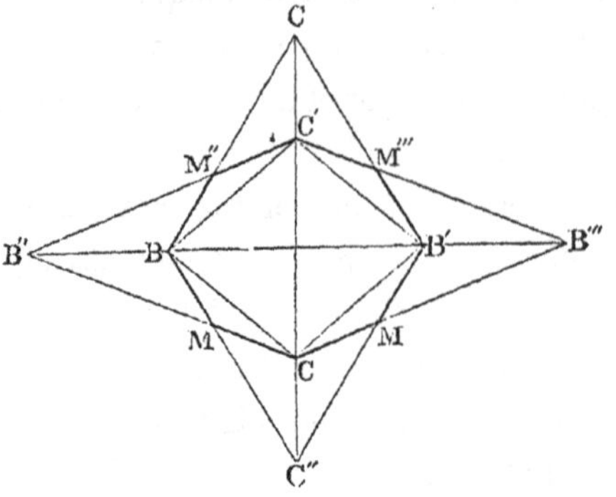

equal to each other. A form constructed upon such a basis cannot be contained under equal and similar planes, or it cannot be a simple form. Like the four-sided pyramid of an oblong rectangular basis, it is considered as an auxiliary or intermediate form, whose resolution gives the simple forms of which it consists.

47. *Resolution of the Auxiliary Form.*

This resolution is executed by enlarging those places which are equal and similar to each other, till the rest disappear. In the present case, the resolution produces two different scalene four-sided pyramids of equal axes, their bases being different amongst themselves, as well as from the bases of the fundamental form. They are termed *scalene four-sided pyramids of dissimilar bases, derived from the fundamental form*, and more particularly that in which the long diagonal of the fundamental form remains unaltered, the *pyramid belonging to the long:* that in which the short diagonal remains unaltered, the *pyramidal belonging to the short diagonal of the fundamental form.*

48. *Designation.*

This important difference between the two forms must be expressed by the designation. The designation $(P+n)^m$ used for the scalene eight-sided pyramids in § 37, if referred to forms in connection with the scalene four-sided pyramids, denotes at the same time both of the pyramids just mentioned. In order to make it applicable to each of them singly, it is now necessary to express their difference in respect to the diagonals of the fundamental forms, which is effected by adding the signs $\smile$ and $-$. Thus $(\breve{P}+n)^m$ represents a scalene four-sided pyramid of dissimilar basis, derived from $P+n$ according to the number m, which belongs to the long diagonal; $(\bar{P}+n)^m$, another pyramid of the same properties which belongs to the short diagonal of P.

49. *Ratios.*

The pyramids just mentioned differ from the members of the series (§ 44), 1*st*, in their axes, which, the axis of P being $=a$, are expressed by $2^n ma$; 2*d*, by the ratio of the diagonals of their bases. Let in P the ratio of the axis to the long and to the short diagonal be that of $a : b : c$, or the ratios of the same lines in $P+n$ that of $2^n a : b : c$; the ratio of the homologous lines will be

$$\text{in } (\breve{P}+n)^m = 2^n ma : b : mc,$$

$$\text{in } (\bar{P}+n)^m = 2^n ma : mb : c,$$

as follows easily from the consideration of a few triangles in the figures required for explaining the derivation. Hence the ratio between the diagonals of the bases in the derived forms, that of the fundamental form remaining constant, depends only upon the number *m*. Accordingly, if *m* be supposed constant, all the pyramids of this kind, derived from scalene four-sided pyramids of similar bases, will likewise, as far as they belong to one or the other diagonal, contain similar bases or transverse sections, the axes of the forms, subservient to the derivation, may be whatever they will.

50. *Double Series.*

If those forms are the members of the above series, each *m* gives rise to *two series* of derived pyramids of a dissimilar basis, viz. one whose members refer to the long, and another whose members refer to the short diagonal of the basis of the fundamental form. From each of those series having constant transverse sections in all their members, it appears that the obliquangular four-sided prisms, limiting them, must likewise be subject to the equality of transverse sections or bases. Both the series stand, therefore, between their limits, thus:

$$P-\infty \ldots (\mathring{P}+n)^m \ldots (\mathring{P}+8)^m$$
$$P-\infty \ldots (P+n)^m \ldots (P+8)^m$$

What has been observed in § 44 in respect to position, may be equally applied here. The values of *m* hitherto most commonly, though not exclusively observed, are = 3, 4, 5. But neither in this case are they the only ones.

51. *Derivations from the Auxiliary Form,* § 41. *Ratios of the Series.*

Instead of the fundamental form P, or of P + *n*, suppose the auxiliary or intermediate form, belonging to it, to take its place; by applying to it the proceeding described in §§ 46, 47, the form produced in a compound one like that in § 46, but its resolution yields again two simple forms, which are also *scalene four-sided pyramids, whose bases are dissimilar to those of the fundamental form.* These pyramids can be distinguished from each other by referring, the one to the long, the other to the short diagonal; from those derived in § 47, by the form to which the derivation is applied, since, in § 47, it is the fundamental form itself, in the present case, the auxiliary form belonging to it; which latter circumstance is expressed in the crystallographic sign by the letter *r* added to P. From the dimensions of the three lines disposed perpendicularly to each other in the pyramid $P+n$, the axis and both the diagonals, being in the ratio of $2^n a : b : c$, it will follow, the ratio of the homologous lines in the scalene four-sided pyramids of dissimilar basis, derived from the auxiliary form, which belongs to the long diagonal of the fundamental form, or in

$$(\breve{P}r+n)^m \text{ to be } = \left(\frac{m+1}{2}\right) . 2^n a : b \left(\frac{m+1}{m-1}\right) c,$$

$$\text{in } (\bar{P}r+n^m \quad = \left(\frac{m+1}{2}\right) . 2^n a : \left(\frac{m+1}{m-1}\right) b : c.$$

The dependence of the ratios between the diagonals of the basis, only from *m*, is evident; and this remains unaltered, how much soever *a* and *n* may be allowed to vary. If the derivation is applied to the whole series in § 44, there will arise two series of scalene four-sided pyramids, which proceed according to the law of the fundamental series, and are limited by obliquangular four-sided prisms, partaking of the similarity in the transverse sections of all the members. The crystallographic designation of the series, between their limits, stands thus:

$$P-\infty \;.\;.\; (\breve{P}r+n)^m \ldots (\breve{\bar{P}}r+|\infty)^m$$

$$P-\infty \ldots (\bar{P}r+n)^m \ldots (\bar{P}r+\infty)^m$$

The values of m are here the same as in the above-mentioned cases. To the position applies what has been said in § 44.

52. *Subordinate Series.*

That there are certain *subordinate* series belonging to the series § 44, which, in reference to the former, may also be considered as the principal series, can easily be supposed from the agreement existing between the forms derived from the scalene, and those derived from the isosceles four-sided pyramid. In order to obtain the members of these subordinate series, enlarge the faces of the fundamental form, fig. 10, on both sides; draw in the enlarged faces

Fig. 10.

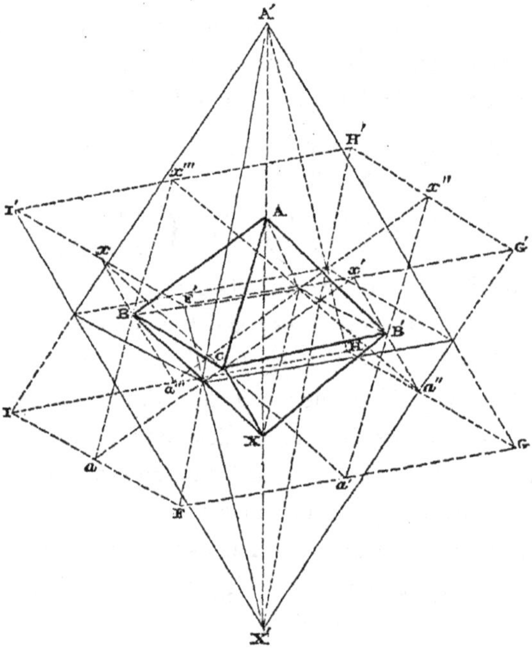

triangles, equal and similar to the faces of the fundamental form; through the point $a, a' \ldots x, x' \ldots$, thus determined, lay rhombs, similar and parallel to the basis of the fundamental form; multiply the axis by a number m; join the points X′, A′, of the lengthened axis by straight lines with the said points $a, a' \ldots x, x'$; and, lastly, reduce the derived pyramid which has a basis similar to that of the fundamental form, to an equal horizontal projection with it. The number m being not yet determined, the coefficient of the subordinate series will be $= \frac{m+1}{2}$; which, if $m+1$ be equal to any power of the number 2, shews the member to which it belongs to be a member of the principal series. The subordinate series, whose coefficient, besides powers of the number 2, contains the factor 3, is termed the *first*, and that which, in the same case, contains the factor 5, the *second* subordinate series. Members of the principal, of the first and of the second subordinate series, which are supposed to belong together, arrange themselves thus: $1 : \frac{3}{4} : \frac{5}{4}$, which numbers express the ratio between their axes. The designation and the position of the members in the subordinate series is evident from the foregoing.

53. *Resolution of the Auxiliary Form*, § 41. *Horizontal Prisms.*

The only form now left for examination is the auxiliary form, § 41; which, as has been mentioned above, is not a simple one, its faces not being homologous. By enlarging only those planes which, amongst themselves, are equal and similar, there arise *two obliquangular four-sided prisms*, unlimited in the direction of their axis, and which, therefore, like all the prisms, are forms in which one dimension is infinite. The axes of the two prisms are perpendicular to each other, and have a horizontal position, if the fundamental form be in the upright position. These axes coincide with the diagonals of the basis of the fundamental form produced to an infinite length. Hence the prisms may be considered as scalene four-sided pyramids, one diagonal of which has become infinite; similarly to the *vertical* prisms, that are scalene four-sided pyramids of an infinite axis. For the sake of distinguishing them from these, they bear the name of *horizontal* prisms, and their general designation is $Pr + n$. More particularly, $\breve{Pr} + n$ expresses the horizontal prism in which the longer diagonal, $\breve{Pr} + \infty$ that in which the shorter diagonal of the fundamental form has remained a finite quantity.

54. *Series of Horizontal Prisms.*

Every scalene four-sided pyramid, whatever its properties may be, has its dependent horizontal prisms. To every series of such forms will, therefore, belong two series of horizontal prisms, to be distinguished by their designation according to the properties of those series. It is, however, remarked, that, if we take the common values of m, as stated above, the angles of several members of those series, and, therefore, the series themselves, will be identical; and, for this reason, one designation will suffice for them. Thus, the number of series of those forms taken together may be reduced to three pairs, of which the first pair belongs to the principal, the others to the subordinate series of scalene four-sided pyramids; each of the diagonals having one series referred to it. The horizontal prisms belonging to a scalene four-sided pyramid are produced by placing tangent planes in its *homologous* terminal edges. If those terminal edges connect the terminal point of the axis with the long diagonal, the prism is said to belong, or to be referred, to the long; if they connect it with the short, it is said to belong, or to be referred, to the short diagonal of P. The horizontal prisms are remarkable forms, often to be met with in the natural combinations.

55. *Limits of the Series of Horizontal Prisms.*

By comparing the horizontal prisms belonging to members of a series of scalene four-sided pyramids, it becomes evident how, whilst the axis of the pyramid increases, the angle of the horizontal prism contiguous to the axis of the fundamental form becomes smaller and smaller; whilst the other, at the intersection of the planes from different apices, becomes greater and greater. If the axis of the pyramid become infinite, or the pyramid itself a vertical prism, the first of these angles will disappear, the second become = 180°. Hence a horizontal prism is transformed, in this case, in a pair of unlimited parallel planes, perpendicular to the diagonals to which they belong. These planes, which may be considered as forms unlimited, as it were, in two directions, are the limits of the series of horizontal prisms, the axis of the pyramid to which they belong being infinite. The opposite limits, or those which appear, if the pyramid have an infinitely small axis, appear as faces perpendicular to the axis. The series of horizontal prisms between their limits may, accordingly, be represented thus:

$$\mathrm{P}-\infty \;\ldots\; \mathrm{P}\breve{\mathrm{r}}+n \;\ldots\; \mathrm{P}\breve{\mathrm{r}}+\infty$$

$$\mathrm{P}-\infty \;\ldots\; \mathrm{P}\bar{\mathrm{r}}+n \;\ldots\; \mathrm{P}\bar{\mathrm{r}}+\infty .$$

For, having the designation in every respect complete, the coefficient of the series to which the horizontal prism belongs must be added to the sign for a particular member,

Although this is not a place for entering into the examination of combined forms, yet that peculiar combination which is produced by the two limits of the series of horizontal prisms, and takes the appearance of a vertical rectangular four-sided prism, not yet limited in the direction of its axis, cannot remain unnoticed. This combination is designated by $P\breve{r} + \infty \;.\; Pr + \infty$; it must carefully be distinguished from the rectangular four-sided prisms, limiting the series of isosceles four-sided pyramids, which, abstraction being made of the face perpendicular to the axis, are simple forms; a property not belonging to the rectangular prism in connection with the scalenc four-sided pyramid.

56. *Derivations from the Hexahedron.*

Besides the rhombohedron, the isosceles, and the scalene four-sided pyramids, there is only one other form existing fit for being considered as a fundamental form, or which, without being derivable from any one of the former, can itself serve as basis to the derivation. This form is the *hexahedron*,

Fig. 11.

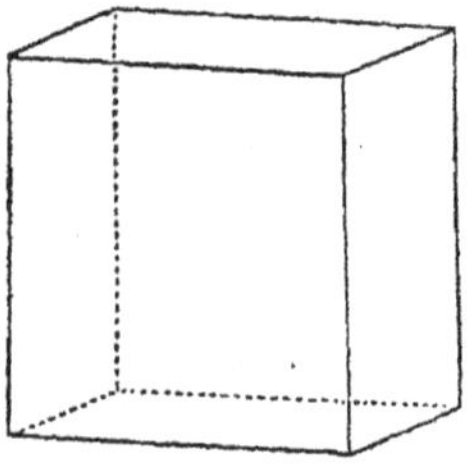

Fig. 12.

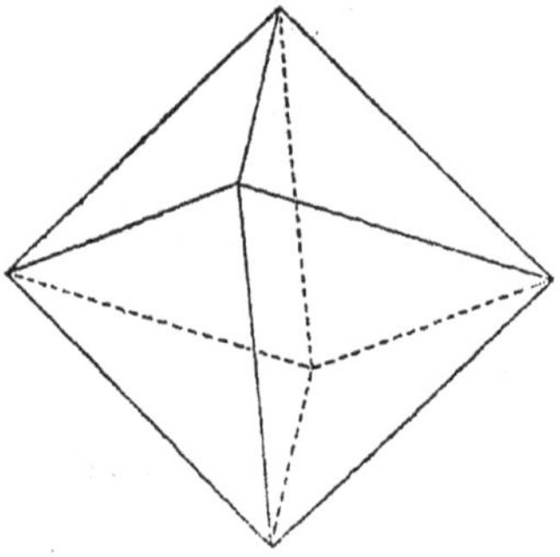

fig. 11. There are, in connection with the hexahedron, a great number of other forms, remarkable for their proper-

ties, differing, however, from each other more by their general aspect than by their dimensions. The complete assemblage of those forms is obtained in the following way. Bring first the hexahedron in an upright position, so as to bring one of its corners uppermost, and the opposite perpendicularly below it, by which means the rhombohedral axis, passing the two corners, becomes vertical; then consider the possible situations of a moveable plane, tangent to the hexahedron in the uppermost point of the said rhombohedral axis. The different situations the plane can effect are seven in number; but in one of them it becomes parallel to the face of the hexahedron itself. Every one of these situations gives rise to a peculiar form; hence there will exist as many forms as the moveable plane can assume different situations, and no more. The forms thus obtained agree in several of their properties, particularly in respect of the kind, number, and situation of their axis, and are as follows:—1. The *hexahedron*, contained under six squares. 2. The *octahedron*, fig. 12, contained under eight equilateral triangles. 3. The *one-edged tetragonal dodecahe-*

Fig. 13.

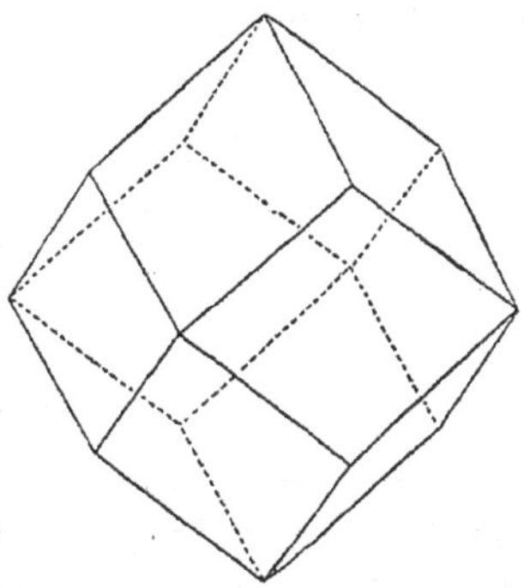

Fig. 14.

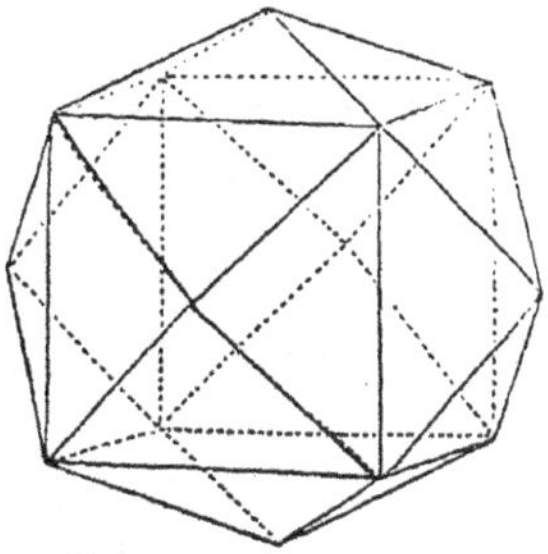

dron, fig. 13, contained under twelve rhombs. 4. The *hexahedral trigonal-icositetrahedron* (fig. 14), contained under twenty-four isosceles triangles, its prominent form being that of the hexahedron. 5. The *octahedral trigonal-icositetrahedron* (fig. 15), contained under twenty-four isosceles triangles, its prominent form being that of the octahedron. 6. The *two-edged tetragonal-icositetrahedron* (fig. 16), contained under twenty-four irregular tetragons,

Fig. 15.

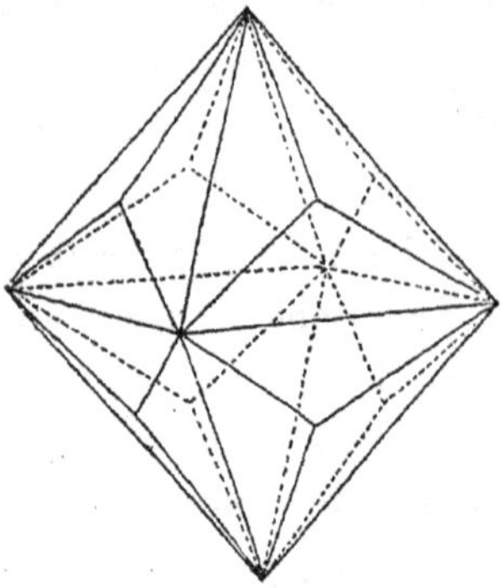

Fig. 16.

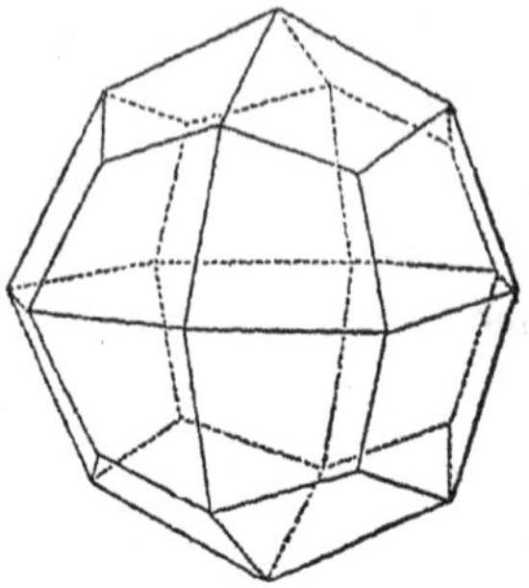

in which two of the opposite angles are equal; and, 7, The *tetracontaoctahedron* (fig. 17), contained under forty-eight scalene triangles. The dimensions of the three first of these forms are constant, a property not to be met with in the others, of which, therefore, several varieties may be found differing from each other by their angles.

57. *Resolution of some of those Forms into Halves.*

This assemblage, though it is complete, seems not to contain all the forms in connection with the hexahedron, as they are produced by nature, such as, for instance, the tetrahedron, the pentagonal dodecahedrons, &c. Nevertheless it contains them, for they arise from a particular resolution of some of those mentioned in § 56. This resolution is effected by enlarging half of the number of planes to be met with in the form to be resolved, according to certain rules, and by making the rest of them disappear. The two forms which arise out of the resolution of one of the former are termed halves (not half forms, as the simple four, six, or eight sided pyramids), and can be easily distinguished by the lesser number of their axes, as some of those that are common to the original forms disappear by the resolution. The two halves are perfectly equal and similar to each other; they differ, however, amongst themselves by

Fig. 17.

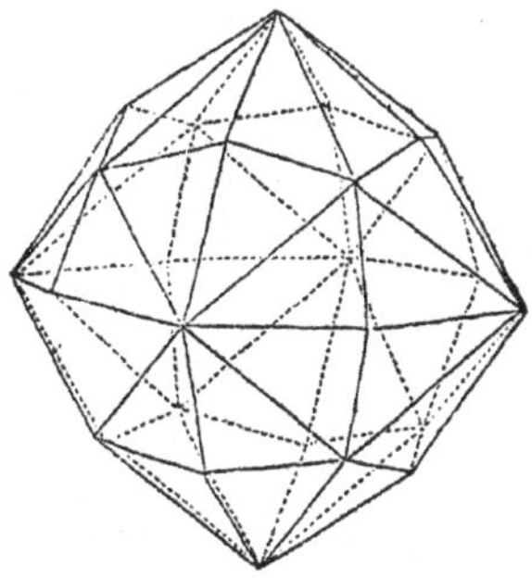

Fig. 18.

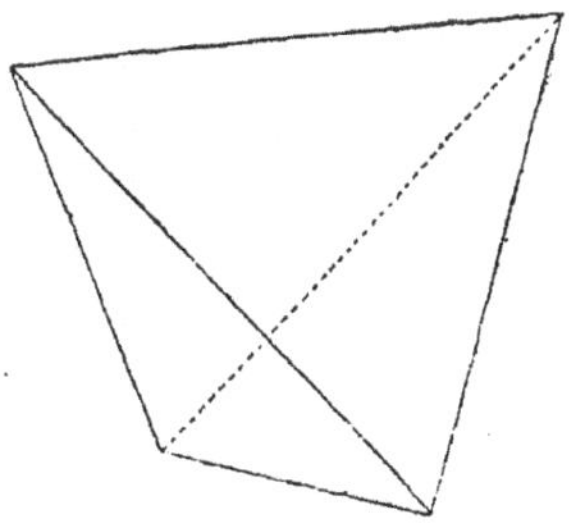

that position in which they are placed by the resolution. The position of one of the halves being considered as the *direct*, that of the other will be the *inverse*. The latter is changed into the former by inverting the vertical rhombohedral axis of the form, so as to bring the upper apex lowermost, and *vice versa*. Thus, the halves produced from the octahedron are the two *tetrahedrons* (fig. 18), contained under four equilateral triangles; those from the hexahedral trigonal-icositetrahedron, the two *hexahedral pentagonal dodecahedrons* (fig. 19), contained under twelve irregular pentagons, which have two pairs of equal angles; those from the octahedral trigonal-icositetrahedron, the *two-edged tetragonal dodecahedrons* (fig. 20), contained under twelve tetragonal faces, in which there is only one pair of equal angles; those from the two-edged tetragonal-icositetrahe-

Fig. 19.

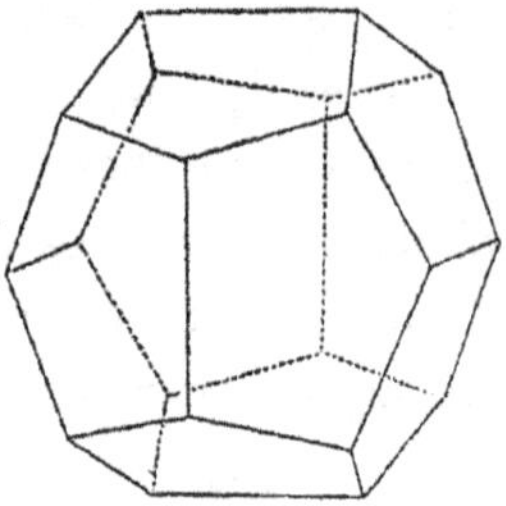

Fig. 20.

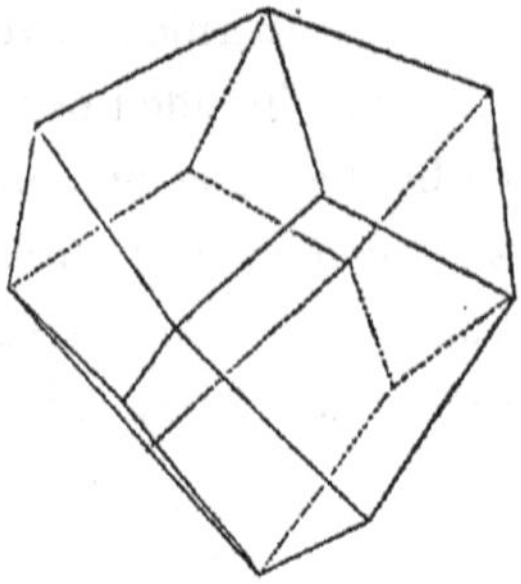

dron, the two *trigonal dodecahedrons* (fig. 21), contained

Fig. 21.

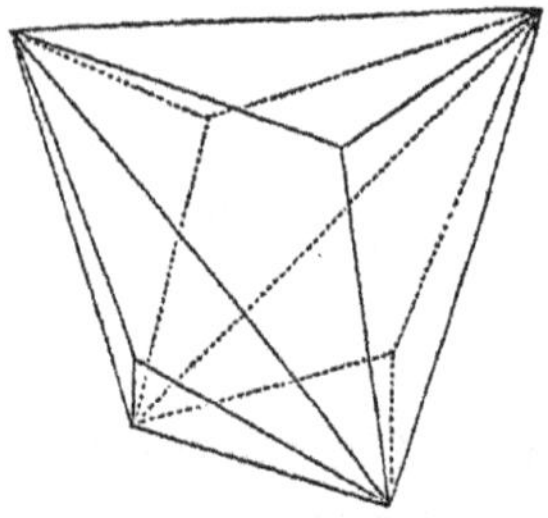

Fig. 22.

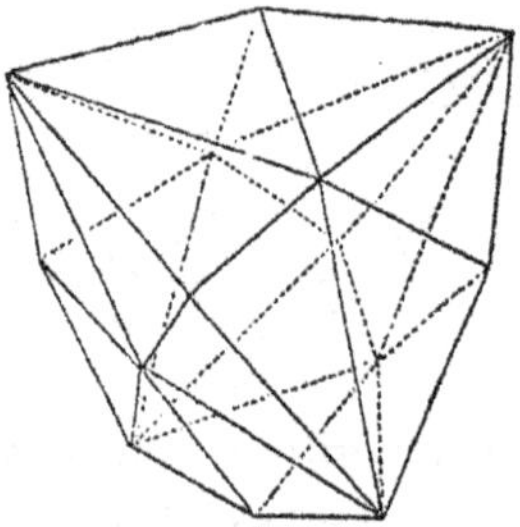

under twelve isosceles triangles; and lastly, the halves of

Fig. 23.

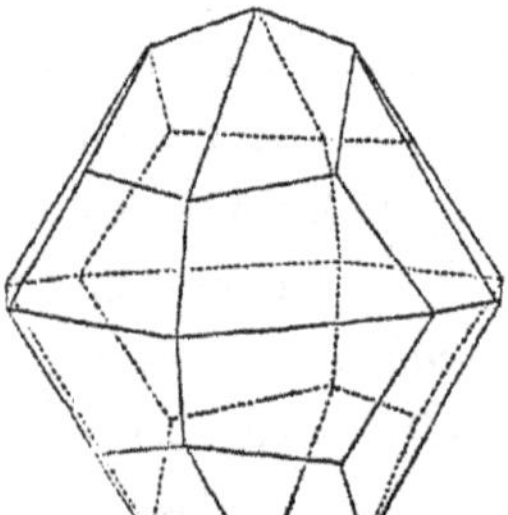

Fig. 24.

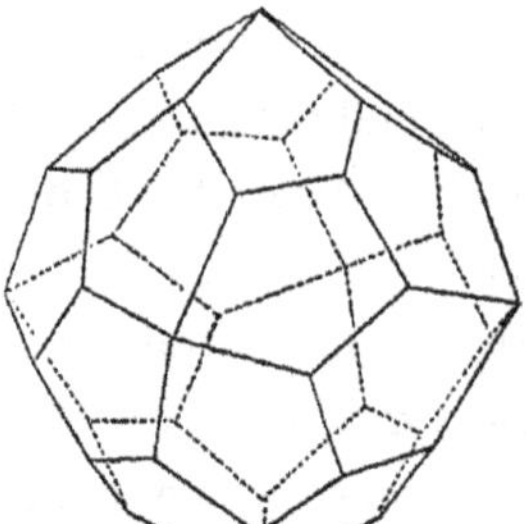

the tetracontaoctahedron, viz. *a*, the *tetrahedral trigonal-*

icositetrahedrons (fig. 22), contained under twenty-four scalene triangles; *b*, the *three-edged tetragonal-icositetrahedrons* (fig. 23), contained under twenty-four irregular tetragonal faces, in which there are not two equal angles; and *c*, the *pentagonal-icositetrahedrons* (fig. 24), contained under twenty-four irregular pentagons, whose angles are all different.

Fig. 25.

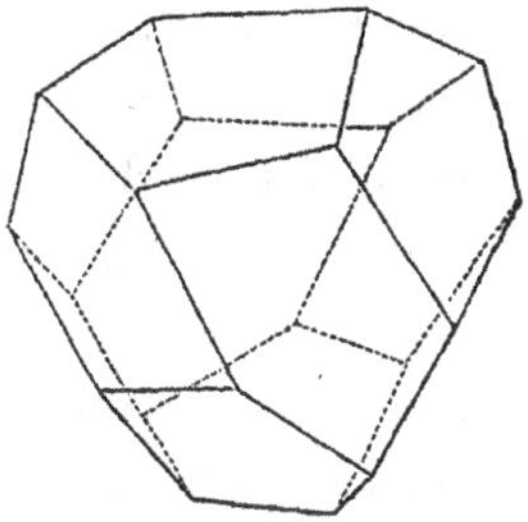

58. *Resolution of some of the Halves into Fourths.*

The three last, or those halves which are obtained from the tetracontaoctahedron, allow of a further dissolution into halves of the halves, or into *fourths*. These are the *tetrahedral pentagonal-dodecahedrons* (fig. 25), contained under twelve irregular pentagons, in which there are not two equal angles. The difference between *right* and *left*, to be met with in this form, and in the pentagonal-icositetrahedron, deserves some attention. If two-fourths are combined in the same position, which they assume in the resolution, the halves are produced; and if the halves, produced by this combination, are combined likewise in their proper position, they will reproduce the original form, out of which the halves as well as the fourths have been obtained by way of resolution. Most of the original forms and their halves are already known in several varieties of minerals; some of the latter, viz. the tetrahedral pentagonal-dodecahedrons, and the pentagonal-icositetrahedrons, have not yet been found.

59. *System of Crystallization and Series of Crystallization.*

If only the *kind* of the fundamental form has been considered, the assemblage of the forms derived from it is called a *system of crystallization;* but if also notice has been taken of its dimensions, the assemblage of the derived forms receives the denomination of a *series of crystallization.* These two notions are of the same kind, and differ only by the number of objects they contain, the latter being a particular determination of the former; and both refer not so much to the mere aggregation of forms, as to the relations prevailing amongst them.

60. *Denomination of the Systems of Crystallization.*

The systems of crystallization, agreeing in number with the fundamental forms, of which there are four, receive their names according to those fundamental forms. That which has been derived from the rhombohedron is called the *rhombohedral system,* because its forms agree with the rhombohedron in their general properties; that which has been derived from the isosceles four-sided pyramid, for the same reason, the *pyramidal system*; that which has been derived from the scalene four-sided pyramid, the *prismatic system,* on account of the great number and variety of prisms it contains; and that which has been derived from the hexahedron, the *tessular system,* in order to intimate that there occurs no other system of crystallization in nature which shares in the general properties of the hexahedron. It is evident that any form, if known, will suffice for the determination of the system to which it belongs, even though this form be a limiting one. Only the right rectangular four-sided prism is an exception, since it may be a simple form in the tessular, a double combination ($P-\infty$. $P+\infty$) in the pyramidal, and a triple combination ($P-\infty$. $\breve{P}r+\infty$. $\bar{P}r+\infty$) in the prismatic system. The consideration of the mere form does not decide at all in this case. Some of the means of removing this uncertainty will be explained hereafter.

61. *Determination of the series of Crystallization.*

A series of crystallization is determined by any one of its members, which is no limiting form, if the dimensions of it are known. These dimensions, if they are not (like those of the hexahedron, the octahedron, &c.) constant, must be made out by direct measurement. From these the dimensions of any other member, or of the fundamental form, if it has not been measured itself, can be found, in compliance with the relations developed above. The series of crystallization proves of great importance for the determination of the natural history species in the mineral kingdom. It is also an external character of the utmost value in the character of the species. This character requires, therefore, the dimensions of a member of the series to be stated, of which the most eligible is the fundamental form. The limits do not determine the series of crystallization, since, in the rhombohedron and in the pyramidal system, they are common to all; in the prismatic, at least to those series which possess a similar transverse section. Hence the obtaining the limiting forms is not sufficient for the character of the natural history species.

62. *General Laws of Combination.*

The second advantage flowing from the above inquiries, consists in the accurate understanding of the qualities of combinations, and of the development of their most general laws. The first of these laws stands thus :—*The forms which nature combines must belong to one series of crystallization.* The second, *The combination must be effected in that position of the several simple forms it contains, which the derivation assigns to them.* Upon these two laws depends the symmetry of the combinations, which, therefore, is not the *fundamental law of crystallization.*

63. *Rhombohedral and Dirhombohedral Combinations.*

An accurate investigation of the combinations is one of the most interesting parts of crystallography. Without entering into minute details, some of their relations may be shortly mentioned, as they convey general ideas of the connection between different forms, and as they are, therefore, of consequence in the system of arrangement and discrimination to be explained in this article.

Combinations of the rhombohedral system, produced by simple forms in such a manner that they appear in the combination with the full number of their faces, and in their proper position, are termed *rhombohedral* combinations. Such are the most common combinations in the rhombohedral system. Suppose, on the contrary, a rhombohedron to combine with itself in a turned position, it will affect the shape of a simple form, and appear as an isosceles six-sided pyramid. It is termed a *dirhombohedron*, and designated by $2(R+n)$. The dirhombohedrons do not arrange themselves with the isosceles six-sided pyramids in the same series (§ 29), because there is a difference existing in the situation of the bases in the one and in the other, as becomes evident in considering the derivation. In a similar way, two equal scalene six-sided pyramids being combined in a turned position, produce a scalene twelve-sided pyramid, which has as little right as the dirhombohedron to be ranked with the simple forms, although its faces are all equal and similar to each other. Combinations produced by, or containing forms of this kind, are termed *dirhombohedral* combinations.

64. *Hemi-rhombohedral and Hemi-dirhombohedral Combinations.*

It sometimes happens in the rhombohedral system, that the forms enter only with half the number of their faces into a combination. If these combinations contain simple forms, they are said to be *hemi-rhombohedral*; if they contain any of the above-mentioned compound forms under this restriction, they are said to be *hemi-dirhombohedral.* A further distinction is to be made among such combinations, in as much as the faces contiguous to one apex are either parallel, or inclined to those of the other. The hemi-rhombohedral, or hemi-dirhombohedral, are said, in the first case, to be *of parallel,* in the other to be *of inclined* faces. An example will put the importance of this difference in its full light: enlarge in a scalene six-sided pyramid the alternating faces contiguous to one apex, and, at the same time, those of the opposite apex, which are *parallel* to the former, the resultant form will take the appearance of a rhombohedron, without in reality being a form of this kind. If, on the contrary, the faces enlarged on the opposite apex are *inclined* to those of the former, the resultant form is contained under six trapezoidal faces.

65. *Pyramidal and Hemi-pyramidal Combinations.*

A relation similar to that which in the rhombohedral system has been expressed by the name of hemi-rhombohedral, is equally found in the pyramidal system, and here it constitutes the *hemi-pyramidal* combinations. The hemi-pyramidal combinations *of parallel* faces refer only to the eight-sided pyramids, which by this affect the shape of isosceles four-sided pyramids, without partaking of their other properties; those *of inclined* faces refer also to the isosceles four-sided pyramids, which yield forms analogous to the tetrahedron, whilst the result of the scalene eight-sided pyramid is a particular form, contained under eight scalene triangles.

66. *Prismatic, Hemi-prismatic, and Tetarto-prismatic Combinations.*

The differences already mentioned are particularly remarkable in the prismatic system. From what has been said in respect to the other systems, it becomes evident what is meant by *prismatic* and by *hemi-prismatic* combinations.. The latter expression refers also to one or to several prisms, whose axes are parallel, if they enter into the combinations with only half the number of their faces. A similar relation of forms in the present system, marked by the expression *tetarto-prismatic*, arises, when of a scalene four-sided pyramid only the fourth part, as to the number, of the faces appear in a combination. Such a combination can also be produced by oblique angular four-sided prisms, whose axes are perpendicular to each other, when only half the number of their faces is visible. The hemi and tetarto-prismatic combinations illustrate the *oblique*, the rectangular as well as the obliquangular prisms produced by nature. These belong altogether to the prismatic system, and none of them is a simple form.

The signs of the hemi-rhombohedral, hemi-pyramidal, and

hemi-prismatic, are composed of the signs of the whole forms, and of the number 2, placed below them like a divisor; instead of the latter, the signs of the tetarto-prismatic combinations contain the number 4 as divisor.

67. *Tessular and Semi-tessular Combinations.*

Those combinations of the tessular system, into which no halves or fourths enter, are called *tessular* combinations, and in the present place they need no further explanation. Combinations of this system, however, containing halves, are termed *semi-tessular* combinations. These allow of a further distinction into semi-tessular combinations *of parallel faces*, and into those *of inclined* faces, as is the case in the hemi-rhombohedral and hemi-pyramidal combinations. The halves peculiar to the former are those in which every face is parallel to another, as the hexahedral pentagonal-dodecahedron, the three-edged tetragonal-icositetrahedron, and the general aspect of the forms undergoes thereby no alteration. The halves peculiar to the latter have not two faces parallel to each other, as the tetrahedron, the tetrahedral trigonal-icositetrahedron, &c., and the combination itself assumes, or at least approaches to a tetrahedral aspect.

Instead of crystallographic signs, *verbal* expressions are employed, in the characteristic part of this article, when describing the forms of the tessular system; because the use of the former is in the present system subject to much greater difficulties, without affording services either so beneficial or so numerous as in the other systems.

68. *Development of the Combinations.*

To develope a combination is to exhibit by itself each of the simple forms which it contains (§ 4). The determination of the kind of these forms has no difficulty whatever. The homologous planes of a combination being enlarged till the rest disappear, one of the forms after the other is produced. To find out their relations, upon the knowledge of which, nevertheless, depends the demonstration of the combinations, is somewhat more circumstantial, and here the series of simple forms, produced by the derivation, shew their great utility. The following paragraph contains a general idea of the mode of proceeding followed for this purpose by crystallography, though its full explanation would require too much calculation to be given in this place.

69. *Line of Combination, General Formulæ for it, and their Application.*

Imagine a combination of two simple forms, whose dimensions are known, and which themselves are in such a position as is peculiar to the system of crystallization to which they belong. Suppose now two terminal edges, or two diagonals, or one terminal edge and one diagonal, that are situated in a plane which, at the same time, passes through the axis, to be prolonged, till they intersect each other above or below, in a finite or infinite distance from a horizontal plane, laid through the centre of the axis. The straight line measuring the distance between the mutual intersection of the two lengthened lines, and that in which one of the lines meets with the horizontal plane, is termed the *line of combination.* The length and situation of the line of combination are evidently dependent upon the dimensions of the combined forms; and inversely the situation of the edge of combination (§ 3.) is dependent upon the length and situation of the line of combination: for, join by a straight line the terminal point of the line of combination with one of the points of intersection between the edges and diagonals of the combined forms, amongst themselves or with each other, the prolongation of this line will equally pass through the other point of intersection, and therefore be identical with the edge of combination. Hence if two forms of a system, in a determined position, produce with a third form lines of combination equal both in length and situation, the edges of combination thus produced will be equally situated; and if the two forms enter at once into combination with the third, the edges of combination produced by the first and the third must be parallel to those between the second and the third.

Crystallography developes general formulæ, expressing the quantity of the line of combination respecting the three first systems of crystallization (this not being required for the tessular system), in which due notice is taken of the kind of forms, their position, and the quality of the edges of combination, namely, whether the faces producing them are contiguous to one or to different apices of the simple forms. Of these formulæ, that which answers to the circumstances of a given combination between two known simple forms, is chosen and determined for that particular case, by substituting, instead of the variable quantities, m and n occurring in them, those finite numbers, which depend upon the place of the members in the series, and those factors by which the axes of the rhombohedrons and others have been multiplied. Thus the quantity of the line of combination, expressed either by a terminal edge, or by a diagonal, or by the axis of one or the other form contained in the combination, is found. Another formula must now be chosen, answering to a combination between *one* of these known forms, and *another unknown*, which produces with it edges of combination parallel to those in the combination of the two known forms. The formula must be determined for the case in question, by the substitution of numbers to m and n. This can only be done in respect to the known forms; m and n of the unknown remain unaltered in the expression. This expression, as is evident from the preceding, must be equal to the value of the line of combination found above. These two equal terms give an equation to be resolved for the values of m and n.

If the unknown form contains only n (as for instance it being a rhombohedron, where $m = 1$), it is determined in the manner already mentioned. If, on the contrary, it contains m and n at once, another equation must be established in order to determine the second quantity, which is effected in the manner shewn already. Some experience in this method of developing the combinations will teach how to take advantage of certain circumstances offering themselves, by which very often it is rendered possible to argue immediately upon one of the unknown quantities m or n, or even sometimes upon both, and to determine the relations of several forms, according to such arguments, without being obliged to employ calculations.

Every simple form being thus determined as member of a series, and furnished with its crystallographic sign; the combination containing those simple forms will be designated by writing the signs for the simple forms one after the other, as some instances in the preceding, § 60, have already shewn. As to the order in which the signs succeed each other in designating a combination, there are two distinct methods, each of which may be applied. Either those forms, whose planes are perpendicular to the axis, are set down first, those whose planes are parallel to it, last; the rest following each other according to the greater or less inclination which their faces have toward the axis, as decreasing from 90° to 0° ; or the forms, according to their kind, are collected into series between their limits, and then set down one after the other. The designation

$$\underset{a}{R-1} \,.\, \underset{b}{(P-2)^3} \,.\, \underset{c}{R} \,.\, \underset{d}{\tfrac{5}{4}R} \,.\, \underset{e}{R+1} \,.\, \underset{f}{(P)^3} \,.\, \underset{g}{R+\infty} \,.$$

represents fig. 26, according to the first method,

Fig. 26.

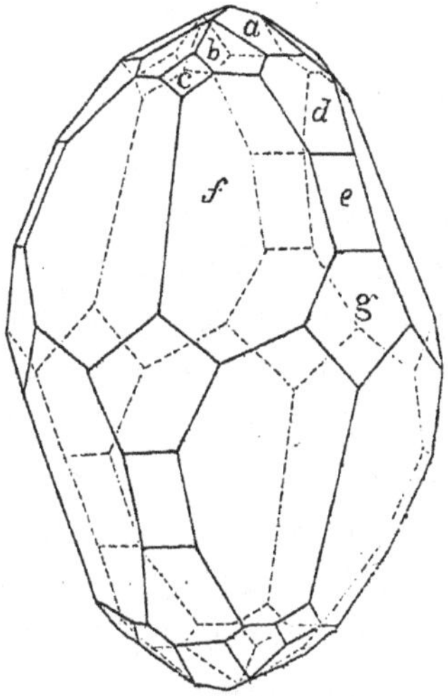

$$\underset{a}{R-1} \,.\, \underset{c}{R} \,.\, \underset{e}{R+1} \,.\, \underset{d}{\tfrac{5}{4}R} \,.\, \underset{g}{R+\infty} \quad \underset{b}{(P-2)^3} \,.\, \underset{f}{(P)^3}$$

the same, after the second method.

70. *Calculation of the Angles at the Edges of Combination.*

After having developed the combination, the last office of crystallography is to calculate the angles of combination, or the incidences of the faces of different forms.

In many instances this angle follows at once from those of the simple forms. If this will not do, crystallography employs general formulæ for the trigonometrical functions of the edge of combination, similar to those for the line of combination, containing equally the variable quantities *m* and *n*. The development of the combinations determines these quantities, and if substituted in the said formulæ, they give the value of a trigonometrical function, commonly of the cosine of the angle, produced by two faces of different forms, meeting in the edge of combination.

II. Cleavage. *Foliated Fracture of Werner.*

Cleavage is the property which minerals possess of splitting, or of being split, in certain determinate directions. The faces or planes thus obtained, which are termed the *faces of the cleavage*, are more or less smooth and shining. The forms contained under these forms are called *forms of cleavage*, or *cleavage-forms*.

The cleavage-forms represent members of the series of crystallization of those mineral species to which the mineral having the cleavage belongs. The same may be inferred of such individuals or species as possess more than one cleavage-form. Hence cleavage extends the application of crystallography in the mineral kingdom, because a mineral, although it be not crystallized, may be cleavable, and thus allow the series, at least the system of crystallization, to be made out by cleavage in most of the cases where no regular crystallizations occur.

The cleavage-forms are designated like those of crystallization. Cleavage-forms, Mohs remarks, if among the the number of those which, in the preceding pages, served as a basis to the derivation, are by preference chosen for representing the fundamental form in the species to which they belong; as in rhombohedral calc-spar, the rhombohedron of 105° 5′. This rule, however, suffers an exception in the case of a rhombohedron or a pyramid, which occur as cleavage-forms, being too obtuse or too acute. For this reason, in pyramidal copper-pyrites, the fundamental form is P, though the cleavage-form is P + 1. No cleavage-form, whose dimensions are infinite, can be employed as a fundamental form.

A few technical expressions are employed by Mohs in determining the cleavage, of which the following may be enumerated. Cleavage, in the three first systems of crystallization, which are those of variable dimensions, is said to be *axotomous*,[1] when it appears as a single plane or face perpendicular to the axis; it is said to be *paratomous*,[2] if it is parallel to the faces of a finite form; and *peritomous*,[3] if it takes place parallel to the axis, and equally distinct in more than one direction. In the prismatic system, the expression *prismatoidal* means a single cleavage face parallel to the axis; and *diprismatic* denotes the cleavage to be parallel at the same time to the faces of a vertical, and to those of a horizontal obliquangular four-sided prism. The expression *monotomous*, referring to the three systems, is more general than any of the foregoing; it comprises the axotomous and the prismatoidal, and is applied where a single cleavage-face is met with, whose position in regard to the axes, that is to say, whether it is parallel or perpendicu-

[1] From αξων, the axis, and τεμνω, I cut; cleavable in one direction.

[2] From παρα, beside, and τέμνω, I cut; with cleavage planes, which are parallel with the planes of the fundamental figure, or are inclined to the axis.

[3] From περί, round about, τεμνω, I cut; with surrounding cleavage planes parallel to the axis.

lar to the axis, has not been determined. The perfection or distinctness of the cleavage, which is next to be considered, refers to the aspect of the faces of the cleavage, viz. whether these are more or less smooth or shining, whether they are streaked, &c. The most important object to be remarked in this respect, is the sameness of quality existing in the faces belonging to one and the same, and the diversity of faces belonging to different cleavage-forms; the latter is a particular illustrated and confirmed by the derivations of the prismatic system. The nature of the faces of the cleavage is one of the means to remove the uncertainty mentioned in § 60, on account of the right rectangular prism. Lastly, it may be remarked, that we must be careful not to confound certain faces of composition (§ 86), which sometimes likewise keep constant directions with the cleavage-forms.

III. Hardness

Is the resistance which solid minerals oppose, when we attempt to separate their parts.

This important character has been rendered more available by Professor Mohs, who gives the following as a useful scale of hardness: the figures 1, 2, 3, &c. are the degrees of the scale.

1. *Talc*, of a white or greenish colour.

2. *Rock-salt*, a pure cleavable variety, or semitransparent *uncrystallized gypsum*, the transparent and crystallized varieties being generally too soft.

3. *Calcareous spar*, a cleavable variety.

4. *Fluor-spar*, in which the cleavage is distinct.

5. *Apatite*, the asparagus stone, or phosphate of lime, from Saltzburg.

6. *Adularia felspar*, any cleavable variety.

7. *Rock-crystal*, a transparent variety.

8 *Prismatic topaz*, any simple variety.

9. *Corundum* from India, which affords smooth cleavage surfaces.

10. The *Diamond*.

If we wish, by means of this scale, to ascertain the hardness of a mineral, we first try which of the members of the scale is scratched by it. In order to save the specimens of the scale it is desirable to begin, in our determination of hardness, with the highest members of the scale, and proceed downwards, until we reach the one which is scratched. We next take a fine hard file, and draw along its surface, with the least possible force, the specimen we are examining, and also one of the minerals in the scale whose hardness is immediately above that of the mineral which has been scratched, in order to compare them together. From the resistance they offer to the file, from the noise occasioned by their passing along it, and from the quantity of powder left on its surface, their relative hardness is deduced. If the two specimens afford the same degree of noise and resistance, and same quantity of powder, their hardness is the same. When, after repeated trials, we are satisfied to which member of the scale of hardness our mineral is most nearly allied, we say its hardness (suppose it to be felspar) is equal to 6, and write after it H = 6.0. If the mineral do not exactly correspond with any degree of the scale, but is found to be between two of them, we say H = 6.5.

In our experiments we must be careful to employ specimens, each of which nearly agree in form and size, and also as much as possible in the shape of their angles. They should also be quite pure, as, if impure, their hardness cannot be determined with accuracy.

IV. Specific Gravity.

The specific gravity of minerals, a very important distinctive character, is determined by means of the hydrostatic balance, the hydrometer or areometer, and Adie's apparatus, a new instrument. (*Vide* Memoirs of the Wernerian Society.) These instruments, and the mode of taking specific gravities, are given under other articles in this work.

II.

CHARACTERS EMPLOYED IN THE DESCRIPTION OF THE SPECIES, SUB-SPECIES, KINDS, AND VARIETIES OF MINERALS.

1. Colour.—2. Common and Particular External Forms.—3. Distinct Concretions.—4. Surface.—5. Lustre.—6. Fracture.—7. Shape of Fragments.—8. Transparency.—9. Opalescence.—10. Streak.—11. Soiling.—12. Tenacity.—13. Frangibility.—14. Flexibility.—15. Adhesion to the Tongue.—16. Unctuosity.—17. Smell.—18. Taste.

I. Colour.

The colours in the mineral kingdom are not so numerous as is generally imagined; and even the varieties, although often extremely beautiful, and apparently infinite in number, bear but a small proportion to the vast series that characterize the various productions of the vegetable and animal kingdoms. Werner, who bestowed great attention on this interesting and beautiful character, enumerates *eight principal colours*, viz. white, grey, black, blue, green, yellow, red, and brown. Each of these principal colours exhibits a greater or less number of varieties; the purest variety is named the *characteristic colour*, thus snow-white, the purest colour of the white series, is the characteristic colour. Colours are further divided into *non-metallic*, or those without metallic lustre, and into *metallic*, or those with metallic lustre.

I. *Definitions of the different Varieties of Colours.*

NON-METALLIC COLOURS.

1. WHITE.

1. *Snow-white* is the purest colour, and agrees with that of new-fallen snow. Examples of this colour occur in Carrara marble and common quartz.

2. *Reddish-white* is snow-white with a slight intermixture of red and grey. Example, heavy-spar.

3. *Yellowish-white* is snow-white with very little lemon-yellow and ash-grey. Example, chalk.

4. *Greyish-white* is snow-white mixed with a little ash-grey. Example, quartz.

5. *Greenish-white* is snow-white mixed with a very little emerald-green and ash-grey. Example, amianthus.

6. *Milk-white* is snow-white mixed with a little Berlin blue and ash-grey. The colour of skimmed milk. Example, calcedony

2. GREY.

1. *Bluish-grey* is ash-grey mixed with a little blue. Example, limestone.

2. *Pearl-grey* is pale bluish-grey mixed with a little red. Example, porcelain jasper, and rarely quartz.

3. *Smoke-grey*, or *brownish-grey*, is dark bluish-grey mixed with a little brown. Colour of dense smoke. Example, dark varieties of flint.

4. *Greenish-grey* is ash-grey mixed with a little emerald-green, and has sometimes a faint trace of yellow. Example, clay-slate and whet-slate.

5. *Yellowish-grey* is ash-grey mixed with lemon-yellow and a trace of brown. Example, calcedony.

6. *Ash-grey*, the characteristic colour, is a mixture of black and white. It is the colour of wood-ashes. Example, zoisite, zircon, and slate-clay.

3. BLACK.

1. *Greyish-black* is velvet-black mixed with ash-grey. Example, basalt, Lydian stone, and lucullite.

2. *Velvet-black* is the characteristic colour of this series. It is the colour of black velvet. Example, obsidian and schorl.

3. *Pitch-black*, or *brownish-black*, is velvet-black mixed with a little yellowish-brown. Example, cobalt ochre, bituminous coal, and some varieties of mica.

4. *Greenish-black*, or *raven-black*, is velvet-black mixed with a little greenish-grey. Example, hornblende.

5. *Bluish-black* is velvet-black mixed with a little blue. Is a rare colour; is well marked in the reniform cobalt ochre from Saalfeld, Thuringia, &c.

4. BLUE.

1. *Blackish-blue* is Berlin-blue mixed with much black, and a trace of red. Example, dark varieties of blue copper.

2. *Azure-blue* is Berlin-blue mixed with a little red. Examples, in bright varieties of blue copper and lapis-lazuli.

3. *Violet-blue* is Berlin-blue mixed with much red, and very little black. Examples, amethyst and octahedral fluor-spar.

4. *Lavender-blue* is pale violet-blue mixed with much grey. Examples, lithomarge and porcelain jasper.

5. *Plum-blue* is Berlin-blue with more red than in violet-blue, and a small portion of brown and black. Examples, spinel and octahedral fluor-spar.

6. *Berlin-blue* is the purest or characteristic colour of the series. Examples, sapphire, rock-salt, kyanite.

7. *Smalt-blue* is Berlin-blue, with much white, and a trace of green. Examples, pale-coloured smalt, named eschel, and some varieties of gypsum.

8. *Duck-blue* is a dark blue colour, composed of blue, much green, and a little black. Examples, ceylanite, talc, and corundum.

9. *Indigo-blue* is a deep blue colour, composed of blue, with a considerable portion of black, and a little green. Example, earthy blue iron of Eckardsberg, in Thuringia.

10. *Sky-blue*, the mountain-blue of painters, is a pale blue colour, composed of blue, green, and a little white. It is the colour of a clear sky, hence its name. Examples, lenticular copper, some varieties of fluor-spar, and of blue-spar.

5. GREEN.

1. *Verdigris-green* is emerald-green mixed with much Berlin-blue and a little white. Examples, the variety of felspar named *amazon stone*, and prismatic liriconite.

2. *Celandine-green* is verdigris-green mixed with ash-grey. Examples, green earth, Siberian and Brazilian beryl.

3. *Mountain-green* is emerald-green mixed with much blue, and a little yellowish-grey; or verdigris-green with yellowish-grey. Examples, beryl, aqua-marine topaz.

4. *Leek-green* is emerald-green with bluish-grey and a little brown. Examples, common actynolite and prase.

5. *Emerald-green* is the characteristic or pure unmixed green. Examples, emerald, and some varieties of green malachite.

6. *Apple-green* is emerald-green mixed with a little greyish-white. Example, chrysoprase.

7. *Grass-green* is emerald-green mixed with a little lemon-yellow. Examples, uranite, smaragdite, and some kinds of green malachite.

8. *Blackish-green* is pistachio-green mixed with a considerable portion of black. Examples, augite, and precious serpentine.

9. *Pistachio-green,* the sap-green of painters, is emerald-green mixed with more yellow than in grass-green, and a small portion of brown. Examples, chrysolite and epidote.

10. *Asparagus-green* is pistachio-green mixed with a little greyish-white, or emerald-green mixed with yellow and a little brown. Example, the variety of phosphate of lime named asparagus stone, from Spain and Salzburg.

11. *Olive-green* is grass-green mixed with much brown and a little grey. Examples, garnet, olivenite, pitch-stone, and olivine.

12. *Oil-green* is emerald-green mixed with yellow, brown, and grey; or pistachio-green with much yellow and light ash-grey. Is the colour of olive-oil. Examples, yellow-blende, beryl, and pitchstone.

13. *Siskin-green* is emerald-green mixed with much lemon-yellow and a little white. Examples, uran-mica and some varieties of phosphate of lead.

6. YELLOW.

1. *Sulphur-yellow* is lemon-yellow mixed with much emerald-green and white. Example, native sulphur.

2. *Straw-yellow* is sulphur-yellow mixed with much greyish-white. Examples, some varieties of schorlite and carpholite.

3. *Wax-yellow* is lemon-yellow mixed with reddish-brown and a little ash-grey; or it may be considered as honey-yellow with greyish-white. Examples, opal and yellow lead-spar.

4. *Honey-yellow* is sulphur-yellow mixed with chestnut-brown. Colour of dark varieties of honey. Examples, fluor-spar and beryl.

5. *Lemon-yellow* is the pure unmixed or characteristic colour. It is the colour of the rind of ripe lemons. Example, yellow orpiment.

6. *Ochre-yellow* is lemon-yellow mixed with a considerable quantity of light chestnut-brown. Examples, yellow earth and jasper.

7. *Wine-yellow* is lemon-yellow mixed with a small portion of red and greyish-white. Examples, Saxon and Brazilian topaz, and octahedral fluor-spar.

8. *Cream-yellow*, or *Isabella-yellow*, contains more red and grey than the wine-yellow, and also a little brown. Examples, bole from Strigau and compact limestone.

9. *Orange-yellow* is lemon-yellow with carmine-red. It is the colour of the rind of the ripe orange. Examples, uran-ochre, and some varieties of molybdate of lead.

7. RED.

1. *Aurora*, or *morning-red*, is carmine-red mixed with much lemon-yellow. Example, red orpiment.

2. *Hyacinth-red* is carmine-red mixed with lemon-yellow, and a minute portion of brown; or aurora-red mixed with a minute portion of brown. Examples, hyacinth and dodecahedral garnet.

3. *Tile-red* is hyacinth-red mixed with greyish-white. The colour of tiles or bricks. Examples, fresh burned porcelain jasper, and some varieties of foliated zeolite.

4. *Scarlet-red* is carmine-red mixed with a very little lemon-yellow. Example, light red cinnabar from Wolfstein.

5. *Blood-red* is scarlet-red mixed with a small portion of black. The colour of blood. Example, pyrope.

6. *Flesh-red* is blood-red mixed with greyish-white. Examples, felspar and heavy spar.

7. *Carmine-red* is the characteristic colour. The colour of carmine. Examples, spinel, particularly in thin splinters, also in rubellite.

8. *Cochineal-red* is carmine-red mixed with bluish-grey. Examples, cinnabar and certain garnets.

9. *Crimson-red* is carmine-red mixed with a considerable portion of blue. Examples, oriental ruby and cobalt bloom.

10. *Columbine-red* is carmine-red with more blue than the preceding variety, and, what is characteristic for this colour, a little black. Example, precious garnet.

11. *Rose-red* is cochineal-red mixed with white. Examples, red manganese and rose-quartz.

12. *Peach-blossom-red* is crimson-red mixed with white. The colour of the blossoms of the peach. Example, red cobalt-ochre.

13. *Cherry-red* is crimson-red mixed with a considerable portion of brownish-black. Examples, spinel, red antimony, and precious garnet.

14. *Brownish-red* is blood-red mixed with brown. The colour of reddle, a ferruginous clay used in drawing, and of columnar clay ironstone.

8. BROWN.

1. *Reddish-brown* is chestnut-brown mixed with a little red and yellow; or chestnut-brown with a small portion of aurora-red. Examples, brown-blende from the Hartz, and pyramidal zircon.

2. *Clove-brown* is chestnut-brown mixed with cochineal-red and a little black. It is the colour of the clove. Examples, rock-crystal and axinite.

3. *Hair-brown* is clove-brown mixed with ash-grey. Examples, wood-opal and brown iron-ore.

4. *Broccoli-brown* is chestnut-brown mixed with much blue and a small portion of green and red. Example, zircon.

5. *Chestnut-brown* is the characteristic or pure brown colour. Example, Egyptian jasper.

6. *Yellowish-brown* is chestnut-brown mixed with a considerable portion of lemon-yellow. Examples, iron-flint and jasper.

7. *Pinchbeck-brown* is yellowish-brown with metallic or semi-metallic lustre. Rather the colour of tarnished pinchbeck. Example, mica.

8. *Wood-brown* is yellowish-brown mixed with much pale ash-grey. Examples, mountain wood and bituminous wood.

9. *Liver-brown* is chestnut-brown mixed with olive-green and ash-grey. The colour of boiled liver. Example, common jasper.

10. *Blackish-brown* is chestnut-brown mixed with black. Examples, mineral pitch from Neufchatel, and moor coal.

Metallic Colours.

1. WHITE.

1. *Silver-white* is yellowish-white, with metallic lustre. Examples, arsenical pyrites and native silver.

2. *Tin-white* is milk-white, with metallic lustre. Example, native antimony.

2. GREY.

1. *Lead-grey* is bluish-grey, with metallic lustre. Example, galena or lead-glance.

2. *Steel-grey* is dark ash-grey with metallic lustre. Example, native platina.

3. BLACK.

1. *Iron-black* is greyish-black with metallic lustre. Example, black or magnetic iron-ore.

4. YELLOW.

1. *Brass-yellow* is sulphur-yellow with metallic lustre. Example, copper-pyrites.

2. *Bronze-yellow* is brass-yellow mixed with steel-grey, and a minute portion of reddish-brown. Example, iron pyrites.

3. *Gold-yellow* is lemon-yellow with metallic lustre. Example, gold-yellow native gold.

5. RED.

1. *Copper-red* is flesh-red with metallic colour. Examples, native copper and copper nickel.

II. *The Play of the Colours.*

If we look on a mineral which possesses this property, we observe, on turning it slowly, besides its common colours, many others, which are distributed in small shining points. It is beautifully seen in the diamond when cut, and also in precious opal, both in the rough state and when polished.

III. *The Changeability of the Colours.*

When the surface of a mineral, which we turn in different directions, exhibits, besides its common colours, different bright colours, that do not change so rapidly, are fewer in number, and occur in larger patches than in the play of the colour, it is said to exhibit what is called the changeability of the colours. The changeability of colour is seen only in particular directions connected with the structure of the mineral, while the play of colour is visible in all directions. Example, Labrador felspar.

IV. *Opalescence.*

Consists in a kind of reflected milky-light, which some minerals exhibit when cut *en cabachon*, or upon plain faces both natural and artificial. This appearance in catseye stone is alleged to be owing to the fibrous structure of the mineral; while in adularia and sapphire, in which it is beautifully seen, it is traced to the foliated structure.

V. *The Iridescence.*

When a mineral exhibits the colours of the prism or the rainbow, arranged in parallel, and sometimes in variously curved layers, it is said to be iridescent. This appearance is supposed in some cases to depend on fissures in the interior of the mineral, as in iridescent rock-crystal, or on the lamellary structure, as in iridescent or rainbow calcedony.

VI. *Tarnished Colours.*

A mineral is said to be tarnished when it shews on its external surface, or on that of the distinct concretions, fixed colours different from those on its interior or fresh fracture.

Tarnished colours are *simple* or *variegated.*

I. *Simple.*

a. Grey,—white cobalt.

b. Black—native arsenic.

c. Brown,—magnetic pyrites.

d. Reddish,—native bismuth.

II. *Variegated.*

The variegated, or party-coloured, are distinguished according to the colour of their basis. Of these the following are enumerated by Werner.

a. Pavonine, or *peacock-tail tarnish.* This is an assemblage of yellow, green, blue, red, and brown colours, on a yellow ground. The colours are nearly equal in proportion, and are never precisely distinct, but always pass more or less into one another. Example, copper-pyrites.

b. Iridescent, or *rainbow tarnish.* In this variety the colours are red, blue, green, and yellow, on a grey ground. It is more beautiful and brighter than the preceding. Example, specular iron-ore or iron-glance of Elba.

c. Columbine, or *pigeon-necked tarnish.* The colours are the same as in the preceding, with this difference, that the tints of colour are paler, and the red predominates. Example, native bismuth of Schneeberg.

d. Tempered-steel tarnish. The colours are very pale blue, green, and very little yellow. Example, grey cobalt.

VII. *Colour Suite.*

Where more than one colour occurs in a species, we find that the tints form characteristic groups or suites, peculiar to the species.

VIII. *Dicröism.*

Some minerals when examined by transmitted light in determinate directions, exhibit different colours. This property is called dichröism. Iolite, tourmaline, and mica exhibit this phenomenon. Iolite is blue in the direction of the axis, yellow or grey perpendicular to it; some varieties of tourmaline are nearly opaque in the direction of the axis, while they display several degrees of transparency, and different green, brown, and blue colours in the direction perpendicular to it; several varieties of mica are green in the direction of the axis, but brown perpendicular to it.

II. Common and Particular External Forms.

1. *Common External Shape.*

Common external shapes are those in which there are neither a determinate number of planes meeting under determinate angles, nor any resemblance to known natural or artificial bodies. As they occur more frequently than the other shapes, they are named *common external shapes.*

Six different kinds are enumerated by Werner, which are distinguished according to their relative length, breadth, and thickness, their relative magnitude, and their connections with other minerals. The kinds are *massive, disseminated, in angular pieces, in grains, in plates, and in membranes.*

1. *Massive* is that common external shape which is from the size of a hazel-nut to the greatest magnitude, and whose dimensions, in length, breadth, and thickness, are nearly alike. It occurs imbedded in other minerals. A very common form.

2. *Disseminated* is from the size of a hazel-nut until it is scarcely visible, and its dimensions, in length, breadth, and thickness, are nearly alike. It occurs imbedded.

3. *In angular pieces.* These are distinguished from massive, by their occurring loose and unattached. Of this external shape there are two kinds:

a. Sharp-cornered, as in *quartz*.

b. Blunt-cornered, as in *common opal*.

4. *In grains.* The form in grains bears the same relation to angular pieces, as disseminated does to massive. They are distinguished into *coarse*, *small*, and *fine grains*, and into *roundish*, *angular*, and *flat grains*. Examples, garnet, platina, and gold.

5. *In plates.* This form is distinguished from massive, by its dimensions in breadth and length very much exceeding that in thickness; it is generally formed in veins. Example, red silver.

6. *In membranes* or *flakes.* This shape is distinguished from the former by its thinness, as it never greatly exceeds the thickness of common paper, and the fracture cannot be seen. Examples, iron-pyrites and silver-glance.

II. *Particular External Shape.*

Particular external shapes differ from the common external shapes, in bearing a resemblance to natural or artificial bodies, and in being far more characteristic and varied in their aspect. There are four different sets, entitled, *longish*, *roundish*, *flat*, and *cavernous*.

1. *Longish Particular External Shapes.*

a. Dentiform. Adheres by its thick extremity, becomes gradually thinner, incurvated, and at length terminates in a free point, so that it resembles a canine tooth, whence its name. Example, native silver.

b. Filiform. Adheres by its thicker extremity, and terminates by an almost imperceptible diminution of thickness, and is usually curved in different directions. It is thinner and longer than the dentiform. Example, native silver.

c. Capillary. When the filiform becomes longer and thinner, it forms the capillary. It is generally much entangled, and sometimes the threads are so near each other that it passes into the compact. Example, native silver.

d. Reticulated is composed of many straight threads, that cross each other under 90°, or under 120° and 60°, and thus form a reticulated shape. Example, silver-glance.

e. Dendritic. In this external shape we can observe a trunk, branches, and twigs, which are distinguished from each other by their thickness, the trunk being the thickest. Examples, native copper and native silver.

f. Coralloidal. When two or three branches, having rounded or pointed extremities, proceed from one stem, the coralloidal external shape is formed. There are usually many stems together. Example, coralloidal aragonite.

g. Stalactitic. A mineral is said to possess a stalactitic external shape, when it consists of different straight, more or less lengthened rods, which are thickest at their attachment, and become narrower at their free extremity, which is rounded or pointed. Examples, calc-sinter and calcedony.

h. Cylindrical consists of long, rounded, straight, imperforated, usually parallel rods, which are attached at both extremities, and are generally thicker at the extremities than the middle. The interstices are either empty or filled up with another mineral. Examples, galena or lead-glance.

i. Tubiform consists of long, usually single, perforated tubes, which are longitudinally knotty. Example, calc-sinter.

k. Claviform is the reverse of *stalactitic;* it is composed of club-shaped parallel rods, which adhere by their narrow extremities. Example, compact black manganese-ore.

l. Fruticose. This external shape has the appearance of cauliflower. Examples, calc-sinter and black manganese-ore.

2. *Roundish Particular External Shapes.*

a. Globular. Under this head are comprehended:

- α. *Perfect globular* or *spherical,* as in alum-slate.
- β. *Imperfect globular,* as in calcedony.
- γ. *Ovoidal* or *elliptical.* Examples, rounded masses of quartz in puddingstone.
- δ. *Spheroidal.* When the spherical is compressed the spheroidal is formed. Example, Egyptian jasper.
- ε *Amygdaloidal.* When the ovoidal is compressed in the direction of its length, the amygdaloidal is formed. Example, zeolite.

b. Botryoidal consists of large segments of small balls, which are irregularly heaped together, and thus leave considerable interstices. It resembles grapes, whence its name. Examples, hematite, hard manganese-ore, and calcedony.

c. Reniform consists of small segments of large balls, which are so closely set together that the interstices are very inconsiderable. Example, calcedony.

d. Tuberose. This shape consists of irregular roundish or longish elevations and depressions. Example, flint.

e. Fused-like or *liquiform.* This consists of numerous very flat rounded elevations, which are generally depressed in the middle. Example, lead-glance.

3. *Flat Particular External Shapes.*

a. Specular has on one side, seldom on two opposite sides, a straight smooth shining surface. It occurs in veins. Example, galena or lead-glance.

b. In leaves. This external shape is in the form of thin, irregularly curved, or straight leaves, of uniform thickness, which adhere by one extremity. It occurs frequently in native gold.

4. *Cavernous Particular External Shapes.*

a. Cellular. A mineral is said to be cellular when it is composed of straight or bent tables, which cross each other in such a manner as to form empty spaces or cells. Example, quartz.

b. Impressed. When one mineral shews the impression of any particular or regular external shape of another mineral, the form it exhibits is named impressed. It borders on the cellular shape; it is formed when a newer mineral is deposited over an older, the form of which it assumes and retains, even after the impressing mineral has been destroyed or removed.

c. Perforated consist of long vermicular cavities, which occupy but an inconsiderable portion of the mass, and terminate on the surface in small holes. When the holes become very numerous, it passes into spongiform. Example, bog iron-ore.

d. Corroded. A mineral is said to be corroded when it is traversed by numerous hardly perceptible roundish holes. Example, quartz.

e. Amorphous is composed of numerous roundish and angular parts that form inequalities, between which there are equally irregular hollows. Example, silver-glance or sulphuretted silver.

f. Vesicular. When a mineral has distributed through its substance many single, usually round, elliptical, and spheroidal, also amygdaloidal, or irregularly-shaped cavities, it is said to be vesicular. Example, lava.

III. Distinct Concretions.

Distinct concretions are those parts into which minerals are naturally divided, and which can be separated from one another, without breaking through the solid or fresh part of the mineral. In describing them, we have to attend to the following appearances. 1. Their shape; 2. Their surface; and, 3. Their lustre.

1. *Shape of the Distinct Concretions.*

They are granular, lamellar, prismatic, radiated, and fibrous.

1. *Granular distinct concretions* are those in which the length, breadth, and thickness are nearly alike. Primitive limestone or marble is composed of granular distinct concretions.

2. *Lamellar distinct concretions* are those in which the length and breadth are nearly equal, and much more considerable than the thickness. Lamellar heavy-spar affords a good example of this kind of concretion.

3. *Prismatic distinct concretions* are those in which the length is very considerable, in comparison of the thickness, thus forming irregular prisms. Examples, amethyst and prismatic heavy spar.

4. *Radiated distinct concretions* are those in which the form is thin prismatic, differing from the preceding in being narrower, and in having the form of rays. The *radiated fracture* of Werner belongs to this division. Example, radiated iron-pyrites.

5. *Fibrous distinct concretions* are those in form of fibres. The *fibrous fracture* of Werner is included under this head. Example, actynolite.

2. *Surface of the Distinct Concretions.*

The surface varies considerably; in some, it is *smooth*, as in hematite; in others it is *streaked*, as in schorl, or it is uneven, as in hornblende.

3. *Lustre of the Distinct Concretions.*

Here the varieties of lustre are the same as already enumerated, and therefore require no particular illustration.

IV. Surface.

The following are the varieties of this character:—

1. *Uneven.* This, of all the kinds of external surface, presents the greatest and most irregular elevations and depressions, yet they are not so considerable as to alter the external shape. Example, surface of balls of calcedony.

2. *Granulated.* When the surface appears like shagreen, it is said to be granulated.

3. *Rough.* This kind of surface is marked with small, scarcely-visible elevations, which we can hardly discover

but by the feel. It has little or no lustre. Example, rolled pieces of common quartz.

4. *Smooth.* Here there is no perceptible inequality, and the surface reflects more light than the preceding kinds of external surface. Example, fluor-spar.

5. *Streaked.* This kind of surface is marked with line-like elevations. It is either *simply streaked* or *doubly streaked.*

a. *Simply streaked* when the line-like elevations run but in one direction.

- α. *Longitudinally streaked.* When the streaks are parallel with the length of the lateral planes. Example, topaz.
- β. *Transversely streaked.* When the streaks are parallel with the breadth of the lateral planes. Example, rock-crystal.
- γ. *Diagonally streaked.* When the streaks are parallel with the diagonal of the planes. Example, garnet.
- δ. *Alternately streaked.* When transverse and longitudinal streaks occur on alternate planes. Example, cubic iron-pyrites.

b. *Doubly streaked,* when the streaks run in different directions. This is of two kinds.

- α. *Plumiformly.* When the streaks run obliquely towards a principal streak, like the disposition of the parts of a feather. Example, plumose native bismuth.
- β. *Reticularly.* When the streaks either cross each other in a promiscuous manner, or under right angles, forming a kind of flat net-work. Example, silver-white cobalt.

c. *Drusy.* When a crystal is coated with a number of minute crystals of the same kind as the mineral itself, the new surface is denominated drusy. Example, common iron-pyrites.

V. Lustre.

Here we have to consider the *intensity* and the *sort* of lustre.

1. *The Intensity of the Lustre.*

Of this there are five different degrees.

1. *Splendent.* A fossil is said to be splendent, when in full day-light its lustre is visible at a great distance, or when the faces produce distinct and well-defined images. Example, galena or lead-glance.

2. *Shining.* When a mineral at a distance reflects but a weak light, it is said to be shining; or when the lustre, although considerable, does not form a distinct image. Example, heavy-spar.

3. *Glistening.* This degree of lustre is only observable when the mineral is near us, and at no greater distance than arms-length; or when the surface no longer forms an image, but reflects a more feeble light than the preceding. Example, porcelain jasper.

4. *Glimmering.* If the surface of a mineral, when held near to the eye in full and clear day-light, presents a number of small, faintly-shining points, it is said to be glimmering. Example, red hematite.

5. *Dull.* When a mineral is destitute of lustre, it is said to be dull. Example, chalk.

2. *The Sort of Lustre.*

The following are the different kinds of lustre:—

1. *Metallic lustre,* so named from its occurring in native metals, is always combined with opacity. It is divided into

perfect and *imperfect.* The perfect metallic lustre occurs in native metals, and in all the species of the orders pyrites and glance; the imperfect or semi-metallic in tantalum ore, uranium-ore, and glance-coal.

2. *Adamantine.* So named from its occurring in the diamond. There are two varieties of this lustre, viz. the *perfect* and *metallic:* the diamond, in its rough and polished states, bright red varieties of ruby silver ore, and some varieties of zinc-blende, and white lead-spar or carbonate of lead, afford examples of the perfect, as the dark varieties of red silver-ore, and of zinc blende, of the imperfect.

3. *Pearly*, so named from its resemblance to the lustre observed in pearls, is divided into *common* and *metallic.* Mica and zeolite are examples of the first, and bronzite of the second.

4. *Resinous* is that which a body would present if smeared with oil. Example, pitchstone and vesuvian.

5. *Vitreous* or *glassy.* The lustre of glass. Example, rock-crystal.

VI. Fracture.

Fracture surfaces are those produced on breaking a mineral. The following are the principal kinds:

1. *Splintery.* When, on a nearly even fracture surface, small wedge-shaped or scaly parts are to be observed, which adhere by their thicker ends, and allow light to pass through, we say that the fracture is splintery. It sometimes passes into even. Example, quartz.

2. *Even* is that kind of fracture in which the surface is nearly even, or without inequalities. Example, Lydian-stone.

3. *Conchoidal* is composed of concave and convex

roundish depressions and elevations, which are more or less regular. When regular, they are accompanied with concentric ridges, as in many shells. Examples, obsidian and rock-crystal.

4. *Uneven*. In this kind of fracture the surface is marked with numerous angular elevations. These inequalities are termed the *grain*, so that we have *coarse* and *fine-grained* uneven fractures. Example, copper-pyrites.

5. *Earthy*. When the fracture-surface shews a great number of very small rather blunt elevations and depressions, which make it appear rough, it is called earthy. Example, chalk.

6. *Hackly*. When the fracture-surface consists of numerous small slightly-bent sharp inequalities, it is said to be hackly. Example, native copper.

7. *Slaty*. In this kind of fracture the mineral splits into plates or slates which are more or less perfect. Example, common roof-slate.

VII. The Shape of the Fragments.

Fragments are those shapes which are formed when a mineral is so forcibly struck, that masses having surrounding cleavage-faces or fracture-surfaces, are separated from it.

The fragments are either *regular* or *irregular*.

I. *Regular fragments* are enclosed in a certain number of regular faces, that meet under determinate angles. The following are the varieties of regular fragments :—

1. *Cubic*, which occur in minerals possessing a rectangular three-fold cleavage, as galena or lead-glance, and rock-salt.

2. *Rhombohedral* or *oblique angular*, which occur in minerals having a three-fold rhombohedral cleavage, as calcareous spar.

3. *Tetrahedral*, or *three-sided pyramidal* and *octahedral*, occur in minerals having a four-fold cleavage in which the folia meet under equal angles, as in fluor-spar.

4. *Dodecahedral.* Fragments of this form occur in minerals having a six-fold cleavage. Example, zinc-blende.

II. *Irregular fragments* are such as have no regular form. The following are the different varieties :—

1. *Cuneiform*, in which the breadth and thickness are much less than the length, and gradually and regularly diminish in magnitude from one end to the other. Example, radiated zeolite.

2. *Splintery*, in which the breadth and thickness are less considerable than the length, but without diminution of magnitude from one extremity to the other. Example, asbestus.

3. *Tabular*, in which the breadth and length are more considerable than the thickness. Example, clay-slate.

4. *Indeterminate angular*, in which the length, breadth, and thickness, are in general nearly alike, but the edges differ much in regard to sharpness, which gives rise to the following distinctions:—

a. Very sharp-edged, as in obsidian.
b. Sharp-edged, as in common quartz.
c. Rather sharp-edged, as in basalt.
d. Rather blunt-edged, as in pumice.
e. Blunt-edged, as in gypsum.
f. Very blunt-edged, as in loam.

VIII. Transparency.

The different degrees of this character are ascertained by observing the relative quantity of light transmitted through their substance. The degrees are five.

1. When a mineral, either in thick or thin pieces, allows the rays of light to pass through it so completely, that we can clearly distinguish small objects placed behind it, as written letters, it is said to be transparent. It is either *simply transparent*, that is, when the body seen through it appears single, as in selenite; or *duplicating*, when the body seen through it appears double, as in calcareous-spar.

2. *Semi-transparent.* When objects can be discerned only through a thin piece, and then appear as if seen through a cloud, it is said to be semi-transparent. Example, calcedony.

3. *Translucent.* When the rays of light penetrate into the mineral and illuminate it, but objects placed behind it cannot be observed either through thick or thin pieces, it is said to be translucent. Example, pitch-stone.

4. *Translucent on the edges.* When the angles and edges are illuminated in the same degree as the whole mineral in the immediately preceding variety of transparency, it is said to be translucent on the edges. Example, horn-stone.

5. *Opaque.* When, even on the thinnest edges of a mineral, no light passes through, it is said to be opaque, as in chalk.

IX. Opalescence.

Some minerals, when held in particular directions, reflect from single spots in their interior, a coloured shining lustre, named opalescence.

X. Streak.

When a mineral is scratched with a hard body, as a knife or file, or is drawn across the surface of a piece of porcelain biscuit, either a powder is formed, or the scratched part acquires a higher lustre. Both these appearances are comprehended under the term streak. The lustre is increased in the streak in the malleable metals, in some glances, and in clays. Some minerals retain their colour in the streak, as the glances and haloids; while others, as many ores, all the pyrites tribe, many blendes, &c. acquire a different colour in the streak.

XI. Soiling or Colouring.

When a mineral, taken between the fingers, leaves some of its particles on them, it is said to *soil* or *colour*.

XII. Tenacity.

By tenacity is understood the relative mobility, or the different degrees of cohesion of the particles of minerals. The degrees of tenacity are the following:—

1. *Brittle.* A mineral is said to be brittle, when, on attempting to cut it with a knife, it emits a grating noise, and the particles fly away in the form of dust, and leave a rough surface, which has in general less lustre than the fracture; it cannot therefore be cut into slices. Example, quartz.

2. *Sectile or Mild.* On cutting minerals possessed of this degree of tenacity, the particles lose their connection in a considerable degree, but this takes place without noise, and they do not fly off, but remain on the knife. Example, galena or lead-glance, and some micas.

3. *Ductile.* Minerals possessing this degree of tenacity can be cut into slices with a knife, extended under the hammer, and drawn into wire. Example, native gold.

XIII. Frangibility.

By frangibility is understood the resistance which minerals oppose, when we attempt to break them into pieces or fragments. It must not be confounded with hardness. Quartz is hard, and hornblende comparatively soft; yet the latter is much more difficultly frangible than the former.

XIV. Flexibility.

This term expresses the property possessed by some minerals, of bending without breaking. Flexible minerals are either *elastical flexible*, that is, if when bent they spring back again into their former direction; or *common flexible*, when they can be bent in different directions without breaking, but remain in the direction in which they have been bent, as molybdena, gypsum, talc, asbestus, and all malleable minerals.

XV. Adhesion to the Tongue.

This character occurs only in such minerals as possess the property of absorbing moisture, which causes them to adhere to the tongue. Example, meerschaum.

XVI. Touch.

This character is of comparatively little importance. Some minerals feel *greasy*, others *meagre;* in order to distinguish the different degrees of unctuosity, the following distinctions are employed:—

1. *Very greasy*, as talc and graphite.
2. *Greasy*, as steatite and fuller's earth.
3. *Rather Greasy*, as asbestus and polished serpentine.
4. *Meagre*, those that do not feel greasy, as cobalt.

XVII. Sound.

Some minerals on being struck emit a *ringing* sound, as rock-crystal; others, when the finger is drawn across them, a *grating* sound, as mountain cork; natural amalgam, when pressed between the fingers, affords a *creaking* sound.

XVIII. Smell.

This character is observed under the following circumstances:—

1. *Spontaneously emitted;* in which case it is,
   a. *Bituminous*, as mineral oil and mineral pitch.
   b. *Faintly sulphureous*, as natural sulphur.
2. *After breathing on it*, in which a *clayey-like* smell is produced, as in hornblende and chlorite.
3. *Excited by friction.*
   a. *Urinous*, in stinkstone.
   b. *Sulphureous*, in iron-pyrites.
   c. *Garlick-like*, or *arsenical*, in native arsenic and arsenic-pyrites.
   d. *Empyreumatic*, in quartz.

XIX. Taste.

This character occurs principally in the saline class, for which it is highly characteristic.

The varieties of it are,

1. *Saline*, common salt.
2. *Sweetish astringent*, natural alum and rock butter.
3. *Styptic*, blue and green vitriol.
4. *Bitter*, natural Epsom salt.
5. *Cooling*, nitre.
6. *Alkaline*, natural soda.

7. *Urinous*, natural sal-ammoniac.
8. *Acid*, sulphuric and carbonic acids.

The *magnetical* and *electrical* properties of minerals may occasionally be used in our account of species.

II. *Fluid Minerals.* In describing liquid minerals, we must attend to their colours, degree of liquidity, lustre, transparency, smell, taste, and specific gravity.

III. *Friable Minerals.* The colour, degree of coherence, lustre, soiling, feel, taste, and hardness, and specific gravity, are to be carefully noted in our description of loosely aggregated or friable minerals.

SYSTEM OF ARRANGEMENT
OF
SIMPLE MINERALS.

CLASS I.

Order I.—GAS.

Genus I. Hydrogen Gas.

1. Pure Hydrogen Gas.
2. Empyreumatic or Carburetted Hydrogen Gas.
3. Sulphuretted Hydrogen Gas.
4. Phosphuretted Hydrogen Gas.

Genus II. Atmospheric Air.

1. Pure Atmospheric Air.

Order II.—WATER.

Genus I. Atmospheric Water.

1. Pure Atmospheric Water.

Genus II. Sea Water.

1. Common Sea Water.

Order III.—ACID.

Genus I. Carbonic Acid.

1. Aeriform Carbonic Acid.

Genus II. Muriatic Acid.

1. Aeriform Muriatic Acid.

Genus III. Sulphuric Acid.

1. Aeriform Sulphuric Acid.
2. Liquid Sulphuric Acid.

Genus IV. Boracic Acid.

1. Prismatic Boracic Acid.

Genus V. Arsenic Acid.

1. Octahedral Arsenic Acid.

Order IV.—SALT.

Genus I. Natron.

1. Hemiprismatic Natron.

Genus II. Glauber Salt.

1. Prismatic Glauber Salt.

Genus III. Nitre.

1. Prismatic Nitre.

Genus IV. Rock-Salt.

1. Hexahedral Rock-Salt.

Genus V. Sal-Almoniac.

1. Octahedral Sal-Almoniac.

Genus VI. Vitriol.

1. Green Vitriol, or Sulphate of Iron. Hemiprismatic Vitriol.
2. Blue Vitriol, or Sulphate of Copper. Tetarto-prismatic Vitriol.
3. White Vitriol, or Sulphate of Zinc. Prismatic Vitriol.

Genus VII. Epsom Salt.

1. Prismatic Epsom Salt.

Genus VIII. Alum.

1. Octahedral Alum.

Genus IX. Borax.

1. Prismatic Borax.

Genus X. Glauberite.

1. Prismatic Glauberite.

CLASS II.

Order I.—HALOIDE.

Genus I. Gypsum.

1. Prismatoidal Gypsum, or Common Gypsum.
2. Prismatic Gypsum, or Anhydrite.

Genus II. Cryolite.

1. Prismatic Cryolite.

Genus III. Alum Stone.

1. Rhombohedral Alum Stone.

Genus IV. Fluor.

1. Octahedral Fluor.

Genus V. Apatite.

1. Rhombohedral Apatite.

Genus VI. Limestone.

1. Aragonite, or Prismatic Limestone.
2. Calcareous Spar, or Rhombohedral Limestone.
3. Dolomite, or Macrotypous Limestone.
4. Breunnerite, or Brachytypous Limestone.
5. Ankerite or Paratomous Limestone.

Order II.—BARYTE.

Genus I. Sparry Iron.

1. Rhombohedral Sparry Iron.

Genus II. Red Manganese.

1. Rhombohedral Red Manganese.

Genus III. Calamine.

1. Prismatic Calamine, or Silicate of Zinc.
2. Rhombohedral Calamine or Carbonate of Zinc.

Genus IV. Tungsten, or Scheelium.

1. Pyramidal Tungsten.

Genus V. Baryte.

1. Strontianite, or Peritomous Baryte.
2. Witherite, or Diprismatic Baryte.
3. Heavy-Spar, or Prismatic Baryte.
4. Celestine, or Prismatoidal Baryte.

Genus VI. Lead-Spar.

1. White Lead-Spar, or Carbonate of Lead. Diprismatic Lead-Spar.
2. Green and Brown Lead-Spar, or Phosphate o. Lead. Rhombohedral Lead-Spar.
3. Red Lead-Spar, or Chromate of Lead. Hemiprismatic Lead-Spar.
4. Yellow Lead-Spar, or Molybdate of Lead. Pyramidal Lead-Spar.
5. Sulphate of Lead, or Prismatic Lead-Spar.

Order III.—KERATE.

Genus I. Corneous Silver, or Chloride of Silver.

1. Hexahedral Corneous Silver.

Genus II. Corneous Mercury, or Muriate of Mercury.

1. Pyramidal Corneous Mercury.

Order IV.—MALACHITE.

Genus I. Copper Green.

1. Uncleavable Copper Green, or Chrysocolla.

Genus II. Liriconite.

1. Prismatic Liriconite, or Lenticular Arseniate of Copper.
2. Hexahedral Liriconite, or Cubical Arseniate of Iron.

Genus III. OLIVENITE.

1. Prismatic Olivenite, or Prismatic Arseniate of Copper.
2. Di-Prismatic Olivenite, or Phosphate of Copper.

Genus IV. BLUE MALACHITE, or BLUE COPPER.

1. Prismatic Blue Malachite.
    * *Velvet Blue Copper.*

Genus V. EMERALD MALACHITE.

1. Rhombohedral Emerald Malachite or Dioptase.

Genus VI. GREEN MALACHITE.

1. Prismatic Green Malachite, or Hydrous Phosphate of Copper.
2. Hemiprismatic Green Malachite, or Common Malachite.
3. Prismatoidal Green Malachite, or Muriate of Copper.

Order V.—MICA.

Genus I. COPPER-MICA.

1. Rhombohedral Copper-Mica, or Micaceous Arseniate of Copper.

Genus II. URAN-MICA, or URANITE.

1. Pyramidal Uran-Mica.
    * Uran-Ochre.

Genus III. COBALT-MICA, or RED COBALT.

1. Prismatic Red Cobalt.

* COBALT-OCHRE.

1. Black Cobalt-Ochre.
2. Brown Cobalt-Ochre.
3. Yellow Cobalt-Ochre.

Genus IV. Antimony-Mica, or White Antimony.

1. Prismatic White Antimony.

* Antimony Ochre.

Genus V. Blue Iron, or Iron-Mica.

1. Prismatic Blue Iron.

Genus VI. Graphite.

1. Rhombohedral Graphite.

Genus VII. Talc-Mica.

1. Prismatic Talc-Mica, or Talc.

1. *Native Magnesia, or Hydrate of Magnesia.* 2. *Ophite, or Precious Serpentine.* 3. *Nephrite.* 4. *Steatite, or Soapstone.* 5. *Figurestone or Algalmatolite.* 6. *Magnesite.* 7. *Meerschaum.* 8. *Lithomarge.* 9. *Mountain Soap.* 10. *Bole.*

2. Rhombohedral Talc-Mica.

3. Hemiprismatic Talc-Mica, or Common Mica.

Genus VIII. Pearl-Mica.

1. Rhombohedral Pearl-Mica.

Order VI.—SPAR.

Genus I. Schiller-Spar.

1. Common Schiller-Spar, or Diatomous Schiller-Spar.
2. Bronzite, or Hemiprismatic Schiller-Spar.
3. Hypersthene, or Prismatoidal Schiller-Spar.
4. Anthophyllite, or Prismatic Schiller-Spar.

Genus II. Kyanite.

1. Prismatic Kyanite.

Genus III. Spodumene.

1. Prismatic Spodumene.

Genus IV. Prehnite.

1. Axotomous Prehnite.

Genus V. Datolite.

1. Prismatic Datolite.

Genus VI. Zeolite.

1. Trapezoidal Zeolite, or Leucite.
2. Dodecahedral Zeolite, or Lapis Lazuli.
3. Hexahedral Zeolite, or Analcime.
4. Paratomous Zeolite, or Cross-stone.
5. Rhombohedral Zeolite, or Chabasite.
6. Diatomous Zeolite, or Laumonite.
7. Prismatic Zeolite, or Mesotype.
8. Prismatoidal Zeolite, or Stilbite.
9. Hemiprismatic Zeolite, or Heulandite.
10. Pyramidal Zeolite, or Apophyllite.
11. Macrytopous Zeolite, or Brewsterite.

Genus VII. Petalite.

1. Prismatic Petalite.

Genus VIII. Felspar.

1. Rhombohedral Felspar, or Nepheline.
2. Prismatic Felspar, Common Felspar, or Potash Felspar.
3. Tetarto-Prismatic Felspar, or Soda Felspar.
4. Polychromatic Felspar, or Lime Felspar.
5. Pyramidal Felspar, Scapolite, Meionite.

Genus IX. Augite.

1. Augite. Paratomous Augite.
2. Hornblende. Hemiprismatic Augite.
3. Epidote. Prismatoidal Augite.
4. Wollastonite. Prismatic Augite.

Genus X. Azure-Spar.

1. Prismatic Azure-Spar, or Lazulite.
2. Prismatoidal Azure-Spar, or Blue Spar.

Order VII.—GEM.

Genus I. ANDALUSITE.
1. Prismatic Andalusite.

Genus II. CORUNDUM.
1. Dodecahedral Corundum, or Spinel.
2. Octahedral Corundum, or Automolite.
3. Rhombohedral Corundum, or Sapphire.
4. Prismatic Corundum, or Chrysoberyl.

Genus III. DIAMOND.
1. Octahedral Diamond.

Genus IV. TOPAZ.
1. Prismatic Topaz.

Genus V. EMERALD.
1. Prismatic Emerald, or Euclase.
2. Rhombohedral Emerald.

Genus VI. QUARTZ.
1. Prismatic Quartz, or Iolite.
2. Rhombohedral Quartz.
3. Uncleavable Quartz.
4. Fusible Quartz.

Genus VII. AXINITE.
1. Prismatic Axinite.

Genus VIII. CHRYSOLITE.
1. Prismatic Chrysolite.
2. Hemiprismatic or Chondrodite.

Genus IX. BORACITE.
1. Octahedral Boracite.

Genus X. Tourmaline.

1. Rhombohedral Tourmaline.

Genus IX. Garnet.

1. Pyramidal Garnet, or Vesuvian.
2. Tetrahedral Garnet, or Helvine.
3. Dodecahedral Garnet.
4. Prismatoidal Garnet, or Grenatite.

Genus XII. Zircon.

1. Pyramidal Zircon.

Genus XIII. Gadolinite.

1. Prismatic Gadolinite.

Order VIII.—ORE.

Genus I. Melane Ore.

1. Prismatic Melan-Ore, or Gadolinite.
2. Anorthetic Melan-Ore, or Allanite.

Genus II. Titanium-Ore.

1. Hemiprismatic Titanium-Ore, or Sphene.
2. Octahedral Titanium-Ore, or Pyrochlor.
3. Peritomous Titanium-Ore, or Rutile.
4. Pyramidal Titanium-Ore, or Octahedrite.
5. Prismatic Titanium-Ore, or Brookite.

Genus III. Zinc-Ore.

1. Prismatic Zinc-Ore, or Red Zinc-Ore.

Genus IV. Red Copper-Ore.

1. Octahedral Red Copper-Ore.

Genus V. Tin-Ore.

1. Pyramidal Tin-Ore.

Genus VI. Wolfram-Ore.

1. Prismatic Wolfram-Ore.

Genus VII. TANTALUM-ORE.

1. Prismatic Tantalum-Ore.

Genus VIII. FERGUSONITE.

1. Pyramidal Fergusonite.

Genus IX. URANIUM-ORE.

1. Uncleavable Uranium-Ore.

Genus X. CERIUM-ORE.

1. Rhombohedral Cerium-Ore.

Genus XI. CHROME-ORE.

1. Octahedral Chrome-Ore, or Chromat of Iron.

Genus XII. IRON-ORE.

1. Octahedral Iron-Ore, or Magnetic Iron-Ore.
2. Rhombohedral Iron-Ore, or Red Iron-Ore.
3. Prismatic Iron-Ore, or Brown Iron-Ore.
4. Diprismatic Iron-Ore, or Lievrite.

Genus XIII. MANGANESE-ORE.

1. Pyramidal Manganese-Ore, or Black Manganese-Ore.
2. Brachytypous Manganese-Ore, or Braunite.
3. Uncleavable Manganese-Ore, or Psilomelane.
4. Prismatoidal Manganese-Ore, or Grey Manganese-Ore.
5. Prismatic Manganese-Ore, or Pyrolusite.

Order IX.—NATIVE METALS.

Genus I. ARSENIC.

1. Rhombohedral.

Genus II. TELLURIUM.

1. Rhombohedral.

Genus III. Antimony.

1. Rhombohedral Antimony.
2. Prismatic Antimony, or Antimonial Silver.

Genus IV. Bismuth.

1. Octahedral Bismuth.

Genus V. Mercury.

1. Liquid Native Mercury.
2. Dodecahedral Mercury, or Native Amalgam.

Genus VI. Silver.

1. Hexahedral Silver.

Genus VII. Gold.

1. Hexahedral Gold.

Genus VIII. Iridium.

1. Rhombohedral.

Genus IX. Palladuim.

1. Octahedral.

Genus X. Platina.

1. Hexahedral.

Genus XI. Iron.

1. Octahedral Iron.

Genus XII. Copper.

1. *Octahedral Copper.*

Order X.—PYRITES.

Genus I. Nickel Pyrites, or Copper-Nickel.

1. Prismatic Nickel Pyrites.

Genus II. Arsenic Pyrites.

1. Axotomous Arsenic Pyrites.
2. Prismatic Arsenic Pyrites.

Genus III. Cobalt Pyrites.

1. Hexahedral Cobalt-Pyrites, or Silver-White Cobalt.
2. Octahedral Cobalt-Pyrites, or Tin-White Cobalt.
3. Isometric Cobalt-Pyrites.
4. Eutomous Cobalt-Pyrites, or Nickeliferous Cobalt-Pyrites.

Genus IV. Iron Pyrites.

1. Hexahedral Iron-Pyrites.
2. Prismatic Iron-Pyrites.
3. Rhombohedral Iron-Pyrites, or Magnetic Pyrites.

Genus V. Copper Pyrites.

1. Pyramidal Copper-Pyrites, or Yellow Copper-Pyrites.
2. Octahedral Copper-Pyrites, or Variegated Copper.

Order XI.—GLANCE.

Genus I. Copper-Glance.

1. Tetrahedral Copper-Glance, or Grey and Black Copper.
2. Prismatoidal Copper-Glance.
3. Prismatic Copper-Glance, or Vitreous Copper.
4. Diprismatic Copper-Glance, or Bournonite.

Genus II. Silver-Glance, or Vitreous Silver.

1. Hexahedral Silver-Glance.

Genus III. Galena, or Lead-Glance.

1. Hexahedral Galena, or Lead-Glance.

Genus IV. Tellurium-Glance, or Black Tellurium.

1. Pyramidal Tellurium-Glance.

Genus V. Molybdena, or Molybdena-Glance.

1. Rhombohedral Molybdena-Glance.

Genus VI. Bismuth Glance.

1. Prismatic Bismuth-Glance.
    * Bismuth-Ochre.

Genus VII. Antimony-Glance.

1. Prismatic Antimony-Glance. Schrifterz.
2. Prismatoidal Antimony-Glance, or Grey Antimony.
3. Axotomous Antimony-Glance. Jamesonite.
4. Peritomous Antimony-Glance, or Sulphuret of Silver and Antimony.

Genus VIII. Melane Glance.

1. Prismatic Melane-Glance, Brittle Silver-Glance of Werner.

Order XII.—BLENDE.

Genus I. Manganese-Blende.

1. Hexahedral Manganese-Blende.

Genus II. Zinc-Blende, or Garnet-Blende.

1. Dodecahedral Zinc-Blende.

Genus III. Antimony-Blende, or Purple Antimony.

1. Prismatic Antimony-Blende, or Purple Antimony.

Genus IV. Ruby-Blende.

1. Rhombohedral Ruby-Blende, or Red Silver.
2. Hemiprismatic Ruby-Blende, or Cinnabar.

Order XIII.—SULPHUR.

Genus I. Sulphur.

1. Prismatoidal Sulphur, or Yellow Orpiment.
2. Hemi-Prismatic Sulphur, or Red Orpiment.
3. Prismatic Sulphur, or Common Sulphur.

CLASS III.

Order I.—RESIN.

Genus I. Mellilite, or Honey-Stone.

1. Pyramidal Mellilite, or Honey-Stone.

Genus II. Mineral Resin.

1. Yellow Mineral Resin, or Amber.
2. Black Mineral Resin.

Order II.—COAL.

Genus I. Mineral Coal.

1. Bituminous Mineral-Coal.
2. Glance-Coal.

Mode of determining a Mineral, or of referring it to its place in the System.

The characters principally employed in the formation of what is called the *Specific Character* are the regular crystallizations (including cleavage); the degrees of hardness; and the specific gravity. The first character given in the *Specific Character* is the system of crystallization to which the form and cleavage of the species belongs. Then follows, together with its dimensions (if known), the fundamental form, from which all the other simple and compound forms are derived. In rhombohedrons, that edge which ends in the apex of the axis, that is to say, the terminal edge is given; for instance, in calcareous spar, $R = 105°5'$; in isosceles four-sided pyramids, both edges, first the terminal one, and then that on the base, are mentioned; for instance, in pyramidal zircon, P = 123°19′, 84°20′; and in scalene four-sided pyramids, first, both of the terminal edges, and then that at the basis, are given: thus, in prismatic topaz, P=141°7′, 101°52′, 90°55′. In this system, besides the dimension of the finite forms, those of the infinite ones, or of the limits, are mentioned, as in the last example, P+∞ =124°19′, and so on; which is very convenient, as the cases in which these can be examined occur more frequently than those in which the edges of pyramids can be measured.

After this follows the indication of the general aspect of the combinations, explained in the foregoing §§ 63–67, under the denominations of dirhombohedral, hemi-prismatic, &c. Of the former are mentioned the angles at the edges of combination, of the latter that terminal edge, which is formed by the intersection of the remaining faces of the scalene four-sided pyramid.

With respect to cleavage, the expression "cleavage, R," for instance in rhombohedral calcareous spar, means, that this mineral has its cleavage parallel to the faces of a rhombohedron, similar to the fundamental form of the species; "cleavage P—∞ . P +∞ .[P +∞]" in pyramidal garnet means that this mineral has its cleavage parallel to the faces of two rectangular prisms, and at the same time perpendicular to their axis; "cleavage $\breve{Pr}+\infty$" in prismatic chrysolite, indicates that the cleavage of this mineral passes at the same time through the axis and the short diagonal of the prism P+∞ ; and "cleavage $(\breve{Pr}+\infty)^3$ = 87°42′. $\breve{Pr}+\infty$. $\overline{Pr}+\infty$," expresses, for instance, in paratomous augite, that the individuals of this species can be cleaved, first parallel to the faces of an obliquangular four-sided prism, of the given dimensions; and, secondly, parallel to planes, which pass through the axis and both diagonals of the prism $P + \infty$; or, what comes to the same, parallel to the faces of a rectangular prism.

Characters mutually excluding each other.

If two or more characters, the one of which excludes the other, be co-existent in the character of an order, or of a genus, as in the genus Corundum, "Tessular, rhombohedral, prismatic," the meaning is, that an individual belonging to this genus must be either tessular, or rhombohedral, or prismatic; because only one of these three can take place at the same time. In the specific character this never happens, because all the forms must belong to one system.

Determination of a Mineral in general.

If a mineral is to be determined, first its *form* must be made out, at least so far as to know the system to which it belongs. Then *hardness* and *specific gravity* must be ascertained, and expressed in numbers. It is sufficient, however, to know the latter to one or two decimals. The specific character requires these data; and they are also of use in the characters of the genera, orders, and classes. This being done, the system may be consulted, and this will at the same time point out what other characters are wanting, so that a mere inspection of the mineral, or a very easy experiment, for instance, to try the streak upon a file, or, still better, upon a plate of biscuit porcelain, will be sufficient. Having advanced in this manner to the character of the species, it will, in some instances, be necessary, and in all cases advisable, for the sake of certainty, to have recourse to the dimensions of the forms. This is particularly necessary, if the genus to which the mineral belongs contains several species having forms of the same system, as is the case in the genus Augite. This determination of the dimensions of the forms may be effected by the common goniometer, the differences in the angles being in general so great, that they cannot easily be mistaken, even by the application of that instrument.

It will seldom be necessary to read over the whole of any character of a class, order, genus, or species, excepting that which includes the individual; one character that does not agree sufficing for exclusion. Thus even the characters of the orders, though the longest, will not be found troublesome.

The application of the method will become very easy and expeditious, by taking particular notice of some characters, which may be termed *prominent*. Such are a metallic aspect; a high degree of specific gravity (particularly if the mineral is not metallic), and a high decree of hardness. The observation of these will immediately decide whether an individual can belong to any particular class, order, genus, or species. It is understood, that if it be not thereby excluded, the other characters must next be examined, till either an excluding one be found, or, if not, the individual may be considered as belonging to that class, order, &c. with which it has been compared and found to agree.

Example.—The following example may serve as an illustration:—Let the form of an unknown mineral be a combination of a scalene eight-sided pyramid, of an isosceles four-sided pyramid, and of a rectangular four-sided prism; the cleavage parallel to the faces of two rectangular four-sided prisms, in a diagonal position to each other; form and cleavage, therefore, *pyramidal*, or belonging to the pyramidal system. Let the *hardness* be = 6.5; *specific gravity* = 6.9.

In this case, both hardness and specific gravity are prominent characters, and exclude at once the individual from the first and third, but not from the second class; with the characters of which its other properties also perfectly agree. Hence the individual belongs to the second class.

Comparing the properties of the individual mineral with the characters of the orders in the second class, hardness and specific gravity will be found too great for the order Haloïde; hardness too great for the orders Baryte and Kerate; both of them too great for the orders Malachite and Mica; and specific gravity too great for the orders Spar and Gem. But in the character of the order Ore, both hardness and specific gravity fall between the fixed limits, and cannot exclude the individual from this order. The other parts of this character are now to be taken into consideration. If the lustre of the individual be metallic, its colour must be black, otherwise it cannot belong to the order Ore. But the lustre is not metallic; therefore the colour of the individual is of no consequence; that is, this conditional part of the character does not affect the individual, and consequently cannot determine its place. Since the lustre is not metallic, the individual must exhibit adamantine, or imperfect metallic lustre: the first will be found particularly in the fracture. The next part of the character refers to minerals of a red, yellow, brown, or black streak; and as the individual gives none of these, its streak being white, this part of the character does not come into consideration. Hardness keeps between the limits, as stated in the character of the order Ore. Should it be = 4.5 and less, the streak must be yellow, red, or black; but hardness is = 6.5, therefore the colour of the streak is indifferent. If the hardness be = 6.5 and more, and streak white, then the specific gravity must be = 6.5 and more. Now, this condition exists: hardness is = 6.5, and the streak is white; the specific gravity being = 6.9, which is greater than 6.5. Lastly, the specific gravity keeps within the limits.

As far as respects the individual which is to be determined, all the characters in the characteristic of the order Ore may be divided into two parts. The first part contains those which refer to the individual; the second those which do not; the last afford no decisive distinctions. But with the first all the properties of the mineral agree. These properties agree consequently with the whole character of the order, as far as it is applicable to the individual, and determine it to belong to the order Ore; or, in shorter terms, to be an *ore*.

Beginners may also compare the characters of the remaining orders. Sometimes they find one individual belonging to two orders, in which case there must be evidently a mistake in the comparison, which would perhaps not have been discovered, had they stopt at the first order, which does not exclude it. In the present case, the want of metallic lustre excludes the individual from the orders Metal, Pyrites, and Glance; hardness from the order Blende; and both hardness and specific gravity from the order Sulphur. The individual can, therefore, be nothing else than an ore. The characters of the genera of the order Ore may next be examined.

If we consider again hardness and specific gravity as prominent, the individual will be immediately excluded from the genera Titanium-ore, Zinc-ore, and Copper-ore, but not from the genus Tin-ore. The form of the pyramidal system, and the white streak, shew that it belongs to this genus. From the genus Wolfram-ore, it is excluded by its too great hardness, and too low specific gravity. From the genera Tantalum, Uranium, Cerium, Chrome, Iron, and Manganese-ore, by hardness and specific gravity, both of them being too high; as also by its white streak, which only agrees with that genus from which the individual differs most by its hardness and specific gravity. The form also does not agree with any in these genera, consequently the individual can belong to no other than to the genus *tin-ore*.

This genus contains but one species. The conclusion that the individual must belong to this species might, nevertheless, be erroneous. There could exist a second species of this genus. The dimensions of the form must now be accurately considered. If these coincide with the angles given in the character, the highest degree of certainty that the individual belongs to or is pyramidal tin-ore, will be obtained.

Perfect determination supposes all the Characters to be known.

The perfect determination of an individual depends, as the above example has shewn, upon the possibility of making out correctly those three properties, viz. form, including cleavage, hardness, and specific gravity. In botany it is the same. The characters must be observable, otherwise the determination will be impossible. In mineralogy, the method affords sometimes more; it leads to a correct determination, even if the knowledge of the form remains imperfect. But it will be an useful rule for beginners to occupy themselves at first with the determination of such individuals as present properties which can be easily and fully investigated. The rest will come of itself, when their knowledge of the mineral kingdom, and particularly of the properties of minerals, increases, and when they have by experience acquired the skill to judge properly of form and cleavage, at least, so far as is necessary for the determination of the system of crystallization, even in those cases where form and cleavage are somewhat difficult to be observed. This exercise is recommended to every naturalist who wishes to acquire a satisfactory knowledge of minerals, by means of the method we have here explained.

Immediate and Mediate Determination.

The method of determination, already illustrated by an example, is termed the *immediate* determination, because it is applied directly to the individual mineral which is to be determined, without the aid of one or several other minerals. If, on the contrary, we must employ one or several other minerals for this purpose, we determine it *mediately.* That variety of hornblende, which is known by the name *amianthus,* occurs in crystals so minute, that their form, supposing it to be regular, cannot be determined even by the most powerful magnifying instruments; cleavage is evidently still less observable. These crystals are flexible, like fibres of flax, their hardness, therefore, cannot be estimated. Their surface has so great an extent in respect to their bulk, that they will swim in water, though endowed with a pretty considerable specific gravity, which, therefore, cannot be ascertained. Some varieties, however, may be observed, whose crystals are a little thicker, though in other respects, these varieties perfectly agree with amianthus. These varieties are not flexible, yet they are too frangible to be able to sustain the trial of hardness. Others are still thicker, but the dimensions of their forms cannot be ascertained on account of their minuteness. They sink in water, and scratch prismatoidal gypsum, but they break, if tried upon calcareous spar. By thus proceeding, we come to such varieties, as possessing a discernible form, allow the cleavage to be investigated; we find their specific gravity about three times that of the water, and their hardness between five and six. These can be determined *immediately*, and will be found to belong to the species of hemiprismatic augite. Chalk, agaric-mineral, clay-slate, and a great many other minerals not allowing of an immediate determination, are determined in the same way, and thus nothing escapes in the natural history method, which in any one of the other methods can be an object of determination.—*Vide Edinburgh Philosophical Journal*, where this system was first explained to English mineralogists, for fuller details in regard to the determination of species. Mr Haidinger's translation of Mohs' Treatise is also recommended to the attention of natural historians.

In the following account of simple minerals, we do not intend to describe all the species enumerated by authors, but to confine ourselves chiefly to those which are the most interesting.

CLASS I.

Specific gravity under 3.8. If solid, is sapid. No bituminous smell.

Order I.—GAS.

Sp. gr. = 0.0001,—0.00014. Elastic. Not acid.

Genus I.—Hydrogen Gas.

Evident smell. Sp. gr. = 0.0001,—0.00014.

1. *Pure Hydrogen Gas.*

Specific Character.—Hydrogenous smell. Sp. gr. = 0.00012.

Geognostic and Geographic Situations.—It rises from limestone, and also from the coal formation, not only in Europe, but in other quarters of the globe.

2. *Empyreumatic or Carburetted Hydrogen Gas.*

Specific Character.—Empyreumatic smell. Sp. gr. = 0.0008.

Geognostic and Geographic Situations.—Rises from marshes and from beds of coal (forming the *fire-damp* of miners) in different parts of Great Britain. It is also one of the aëriform substances given out by volcanoes.

3. *Sulphuretted Hydrogen Gas.*

Specific Character.—Smell of putrid eggs. Taste nauseous and bitter. Sp. gr. = 0.00135.

Geognostic and Geographic Situations.—Rises from marshes, sulphureous springs, and volcanoes. It is met with in many places in Great Britain.

4. *Phosphuretted Hydrogen Gas.*

Specific Character.—Smell of putrid fish. Sp. gr. unknown.

Geognostic and Geographic Situations.—It rises from marshy places, where organic substances are in a state of decomposition.

Genus II.—Atmospheric Air.

Without smell or taste. Sp. gr. =0.001,—0.0015.

1. *Pure Atmospheric Air.*

Specific Character.—Without smell or taste.

Constituent Parts.—Azotic gas 79, oxygen gas 21, =100. Besides these, the chief constituents, other matters occur in the atmosphere.

Forms the atmosphere which surrounds the earth.

Order II.—WATER.

Liquid. Tasteless, or with sensible taste and smell. Sp. gr. =1.0,—1.0269.

Genus I.—Atmospheric Water.

Without smell or taste. Sp. gr. = 1·0.

1. *Pure Atmospheric Water.*

Specific Character.—Without smell or taste.

Constituent Parts.—Oxygen 88,94, hydrogen 11.06, =100. *Berzelius.* The purest natural water contains small portions of foreign matter.

This is common rain, river, and spring water. Mineral waters might be introduced into this part of the system.

Genus II.—Sea Water.

Sensible smell and taste. Sp. gr. = 1.0269.

1. *Common Sea Water.*

Specific Character.—Bitter nauseous taste, and disagreeable smell.

Constituent Parts.—In 10,000 parts of water of the Firth of Forth, common salt 220.01, sulphate of soda 33.16, muriate of magnesia 42.09, muriate of lime 7.84, = 303.09.

Order III.—ACID.

Sp. gr. = 0.0015,—3.7. Acid.

Genus I.—Carbonic Acid.

Taste slightly acid. Sp. gr. = 0.0018.

1. *Aeriform Carbonic Acid.*

Specific Character.—Elastic. Taste acidulous and pungent.

Constituent Parts.—Carbon 27.4, oxygen 72.6, = 100. *Berzelius.*

Geognostic and Geographic Situations.—Occurs in marshy places and in acidulous waters, in Great Britain and other countries.

Genus II.—Muriatic Acid.

Smell of saffron, and strong acid taste. Sp. gr. = 0.0023.

1. *Aeriform Muriatic Acid.*

Specific Character.—Expansible. Smell pungent and suffocating.

Constituent Parts.—Muriatic acid 73.31, water 24.69. *Berzelius.*

Geographic Situations.—Rises from volcanoes.

Genus III.—SULPHURIC ACID.

If gaseous, the smell is sulphureous. If liquid, the taste is strongly acid. Sp. gr. = 0.0025,—1.5.

1. *Aeriform Sulphuric Acid.*

Specific Character.—Expansible. Sp. gr. = 0.0028.

Constituent Parts.—Sulphur 50.144, oxygen 49.856, = 100. *Berzelius.*

Geognostic Situation.—Often rises in considerable quantities from volcanoes.

2. *Liquid Sulphuric Acid.*

Specific Character.—Liquid. Sp. gr. = 1.8,—1.9.

Geognostic Situation.—Occurs in volcanic districts in Italy, America, and Java.

Genus IV.—BORACIC ACID.

Solid. Sp. gr. 1.4 to 1.5.

1. *Prismatic Boracic Acid.*

Specific Character.—Prismatic. Pyramid unknown. Occurs in scaly crusts. Taste first sourish, or sub-acid, then bitter and cooling, and lastly sweetish.

Constituent Parts.—Boron 25.83, oxygen 74.17. *Berzelius.*

Geognostic and Geographic Situations.—Found on the edges of hot springs in Italy, &c.

Genus V.—Arsenic Acid.

Solid. Sp. gr. above 3.0.

1. *Octahedral Arsenic Acid.*

Arsenic oxyde.—*Haüy.*

Specific Character.—Tessular. Cleavage octahedral. Taste sweetish-astringent. Hardness =1.5. Sp. gr. =3.6—3.7.

Description.—Colour white, often inclining to yellow. Occurs in delicate capillary crystals; also massive, in crusts, stalactitic, reniform and botryoidal. Lustre vitreous, inclining to adamantine. Translucent or opaque.

Constituent Parts.—Arsenic 75.82, oxygen 24.18, =100. *Berzelius.*

Geognostic and Geographic Situations.—Occurs in veins at Andreasberg in the Hartz.

Order IV.—SALT.

Sp. gr. =1.2,—2.9. Solid. Not acid.

Genus I.—Natron.

Prismatic. Taste pungent and alkaline. Hardness, =1.0,—1.5. Sp. gr. =1.4,—1.6.

1. *Hemiprismatic Natron.*

Hemiprismatisches Natron-Salz, *Mohs.*—Natürliches Mineral-Alkali, *Werner.*—Soudé carbonatée, *Haüy.*

Specific Character.—Hemiprismatic. Cleavage prismatic.

Description.—Its chief colours are grey, white, and yellow. Occurs in acicular crystals, in radiated and granular distinct concretions; also in loose earthy particles, and in crusts. Is more or less translucent.

Constituent Parts.—Dry carbonate of soda 32.6, sulphate of soda, 20.8, muriate of soda, 15.8, water of crystallization 31.6, = 100.0.

Geognostic and Geographic Situations.—Occurs in crusts on rocks and soils of different kinds; and also in the waters of natron lakes and springs. It is particularly abundant in the natron lakes in Egypt.

Genus II.—Glauber Salt.

Prismatic. Taste, first cooling, then saline and bitter. Hardness = 1.5,—2.0. Sp. gr. 1.4,—1.5.

1. *Prismatic Glauber Salt.*

Prismatisches Glauber Salz. *Mohs.*—Natürliches Glauber Salz, *Werner.*—Soudé sulphatée, *Haüy.*

Specific Character.—Prismatic. Pyramid unknown. Combination hemiprismatic. $\frac{\breve{P}}{2}$. Cleavage, $\breve{P}r + \infty$ perfect. Less perfect, $\bar{P}r + \infty$. (Fig. 29, 28.)

Description.—Colour white, sometimes inclining to yellow. Occurs in acicular crystals, granular concretions, stalactitic, in loose earthy particles, and in crusts. More or less translucent.

Constituent Parts.—Sulphuric acid 24.84, soda 19.39, water 55.77, = 100.

Geognostic and Geographic Situations.—Occurs on soils and rocks of different descriptions, in Scotland, England, and other countries.

Genus III.—Nitre.

Prismatic. Taste cooling and saline. Hardness = 2.0. Sp. gr. = 1.9, — 2.0.

1. *Prismatic Nitre.*

Prismatisches Nitrum-Salz, *Mohs.* Natürlicher Salpeter, *Werner.* Potasse nitratée, *Haüy.*

Specific Character.—Prismatic. Pyramid = 132° 22′; 91° 15′; 107° 43′. Cleavage, Pr + ∞ = 120°. More distinct, $\breve{\text{Pr}}$ + ∞. (Fig. 30, 29.)

Description.—Colour white, grey, and sometimes yellow. Occurs in acicular crystals, in crusts, and in fibrous concretions. More or less translucent.

Constituent Parts.—Potash 55.28, nitric acid 44.72.

Geognostic and Geographic Situations.—Occurs incrusting limestone, marl, sandstone, calc-tuff, chalk, and on soils of particular kinds. It is collected in limestone caves in Italy, in caves of various descriptions in America, and in abundance from the surface of the ground in many of the Tartarian plains.

Genus IV.—Rock-Salt.

Tessular. Taste saline. Hardness = 2.0. Sp. gr. = 2.2, — 2.3.

1. *Hexahedral Rock-Salt.*

Hexaedrisches Steinsalz, *Mohs.* Natürlich Kochsalz, *Werner.*

Specific Character.—Tessular. Cleavage, hexahedral.

Description.—Most frequent colours grey and white; sometimes also blue, red, yellow, and green. Occurs in granular, fibrous, radiated, and prismatic concretions; massive, dentiform, and stalactitic. Lustre between vitreous and resinous. More or less transparent and translucent.

Constituent Parts.—Muriate of soda 98.32, sulphate of lime 0.65, muriate of magnesia 0.02, muriate of lime 0.01 undissolved matter 1.0, =99. *Henry.*

Geognostic and Geographic Situations.—It occurs in beds, imbedded masses, and veins, associated with salinife-rous clay, gypsum, limestone, sandstone, and anhydrite, in the *salt formation;* also in layers and crusts on soils of particular kinds, and deposited on the shores of salt lakes, and in the vicinity of salt springs. Occurs abundantly in Cheshire, and also in other parts of England.

Genus V.—Sal Ammoniac.

Tessular. Taste pungent and urinous. Hardness = 1.5, —2.0. Sp. gr. 1.5, — 1.6.

1. *Octahedral Sal Ammoniac.*

Octaedrisches Salmiac, *Mohs.* Natürlicher Salmiac, *Werner.* Ammoniaque Muriatée, *Haüy.*

Specific Character.—Tessular. Cleavage, octahedral.

Description.—Colours white, grey, yellow, and some-times green and bluish. Occurs in granular and fibrous concretions; also in efflorescences, in crusts, stalactitic, bo-tryoidal, tuberose, and corroded. More or less translucent.

Constituent Parts.—Muriate of ammonia 99.4, sulphate of ammonia 0.5. *Klaproth.*

Geognostic and Geographic Situations.—Occurs in crusts in the fissures and on the surfaces of volcanic rocks, as in Vesuvius, Ætna, &c.

Genus VI.—Vitriol.

Pyramidal, prismatic. Taste, astringent. Hardness =2.0, —2.5. Sp. gr. =1.8, —2.3.

1. *Green Vitriol, or Sulphate of Iron. Hemiprismatic Vitriol.*

Hemiprismatisches Vitriol-Salz, *Mohs.* Eisen Vitriol, *Werner.* Fer Sulfatée, *Haüy.*

Specific Character.—Prismatic. Combination hemi-prismatic $\frac{P}{2} = 102° 35'$, $P + \infty = 82° 21'$. Cleavage $= P + \infty$. Inclination of $P - \infty$ to $P + \infty = 99° 23'$. Green. Hardness $= 2.0$. Sp. gr. $= 1.8, -2.9$.

Description.—Colour green. Occurs regularly crystallized, in fibrous concretions, massive, stalactitic, botryoidal, and reniform. More or less translucent and transparent.

Constituent Parts.—Sulphuric acid 31.02, oxide of iron 27.19, water 41.79, = 100. *Kobel.*

Geognostic and Geographic Situations.—Occurs in coal and iron mines, both in Scotland and England.

2. *Blue Vitriol, or Sulphate of Copper. Tetartoprismatic Vitriol.*

Prismatisches Vitriol-Salz, *Mohs.* Kupfervitriol, *Werner.* Cuivre Sulfatée, *Haüy.*

Specific Character.—Prismatic. Combination, tetarto-prismatic. Cleavage, two faces, one more distinct than the other; incidence, 149° 2′. Hardness $= 2.5$. Sp. gr. $= 2.2, -2.3$.

Description.—Colour blue. Occurs distinctly crystallized, massive, stalactitic, and dentiform. More or less transparent and translucent.

Constituent Parts.—Oxide of copper 31.72, sulphuric acid, 31.72, water, 36.14.

Geognostic and Geographic Situations.—Occurs in copper mines in England and Ireland.

3. *White Vitriol, or Sulphate of Zinc. Diprismatic Vitriol.*

PrismatischesVitriol-Salz, *Mohs.* Zink-vitriol, *Werner.* Zink sulfatée, *Haüy.*

Specific Character.—Prismatic. P = 127° 27; 126° 45; 78° 5′. P + ∞ = 90° 42′. Cleavage $\breve{P}r$ + ∞ very perfect. White. Hardness = 2.0, — 2.5. Sp. gr. = 2.0, — 2.1.

Constituent Parts.—Oxide of zinc 27.5, sulphuric acid, 20.0, water 50.0. *Klaproth.*

Geognostic and Geographic Situations.—Occurs in mines where blende is met with, both in Flintshire and Cornwall.

Genus VII.—Epsom Salt.

Prismatic. Taste bitter and saline. Hardness = 2.0, — 2.5. Sp. gr. = 1.7, — 1.8.

1. *Prismatic Epsom Salt.*

Prismatisches Bittersalz, *Mohs.* Natürlicher Bittersalz, *Werner.* Magnesie Sulfatée, *Haüy.*

Specific Character.—Prismatic. P = 127° 22′; 126° 48′; 78° 7′. P + ∞ = 90° 38′. Cleavage Pr + 8, very perfect.

Description.—Colours white and grey. Occurs in crusts, botryoidal, reniform, and crystallized. Transparent or translucent.

Constituent Parts.—Magnesia 16.0, sulphuric acid 32.53. water 51.43.

Geognostic and Geographic Situations.—Occurs along with natural alum at Hurlet near Paisley.

Genus VIII.—Alum.

Tessular. Taste sweetish, astringent. Hardness = 2.0, —2.5. Sp. gr. = 1.7, — 1.8.

1. *Octahedral Alum.*

Octaedrisches Alaun, *Mohs.*—Natürlicher Alaun, *Werner.*—Alumine Sulphatée Alcaline, *Haüy.*

Specific Character.—Tessular. Cleavage octahedral.

Description.—Colours white. Occurs in crystals; but more frequently in farinaceous efflorescences, stalactitic, and in fibrous concretions. Translucent.

Constituent Parts.—Alumina 10.8. Potash 10.1. Sulphuric acid 33.7. Water 45.4.—*Gmelin.*

Geognostic and Geographic Situations.—Generally occurs incrusting aluminous minerals, in which situation it is met with in various parts of Scotland and England.

Genus IX.—Borax.

Borax-Salz, *Mohs.*

Prismatic. Taste feebly sweetish and alkaline. Hardness = 2.0,—2.5. Sp. gr. = 1.7,—1.8.

1. *Prismatic Borax.*

Prismatisches Borax-Salz, *Mohs.*—Soude Boratée, *Haüy.*

Specific Character.—Prismatic. P = 150° 9′; 120° 20′; 67° 3′. P + ∞ = 52° 53′. Combination hemiprismatic, $\frac{P}{2}$ = 120° 23′. Cleavage $(Pr + \infty)^3$ = 88° 8′. More distinct $\breve{P}r + \infty$. (Fig. 32, 29.)

Description.—Colours white, grey, and green. Occurs crystallized, internally shining and resinous. Fracture, flat, conchoidal. Semitransparent.

Constituent Parts.—Soda 16.31. Boracic acid 36.59. Water 47.10.—*Berzelius.*

Geognostic and Geographic Situations.—Occurs in the soil, and in the water of springs in Thibet and Persia.

Genus X.—GLAUBERITE.

Brithyn-Salz, *Mohs.*

Prismatic. Taste feebly saline and astringent. Hardness = 25,—3.0. Sp. gr. 2.75, —2.85.

Prismatic Glauberite.

Prismatisches Brithyn-Salz, *Mohs.*—Glauberite, *Haüy.*

Specific Character.—Prismatic. Pyramid unknown. Combination hemiprismatic. Cleavage $\frac{Pr}{2}$, perfect. Indistinct $\bar{P} + \infty = 104° 28'$. (Fig. 35, 30.)

Description.—Colours white and yellow. Occurs crystallized in oblique four-sided prisms. Shining lustre. Fracture conchoidal. Transparent. Brittle.

Constituent Parts.—Sulphate of lime 49.0. Sulphate of soda 51.0.—*Brongniart.*

Geognostic and Geographic Situations.—Occurs in masses of rock-salt, associated with clay, at Villarubia near Ocanas, in the province of Toledo, in Spain.

CLASS II.

Specific gravity above 1.8. Tasteless.

Order I.—HALOIDE.[1]

Not metallic. Streak white or grey. Hardness = 1.5, —5.0. Sp. gr. = 2.2.—3.3.

[1] From ἅλς, salt; and εἶδος, the appearance (habitus).

If pyramidal or prismatic, the hardness = 4.0, and less. If tessular, the hardness = 4.0. If single highly perfect faces of cleavage, the sp. gr = 2.4, and less. If the hardness is under 2.5, the sp. gr. = 2.4, and less. If the sp. gr. = 2.4, and less, the hardness is under 2.5, and no resinous lustre.

Genus I.—Gypsum.

Prismatic. Hardness = 1.5,—3.5. Sp. gr. = 2.2,—3.0. If the sp. gr. is above 2.5, there are cleavages in three directions, perpendicular to each other, and of which two are more distinct than the other.

1. *Prismatoidal Gypsum*, or *Common Gypsum*.

Prismatoidisches Gyps-Haloid, *Mohs.*—Chaux sulfatée, *Haüy.*

Specific Character.—Prismatic. Pyramid = 149° 33′; 135° 32′; 54° 52′. $P + \infty = 110° 30′$. Combination hemiprismatic, $\frac{P}{2} = 149° 33′$. Cleavage, $\overset{\circ}{Pr} + \infty$, very perfect and distinct, $\bar{P}r$, $\bar{P}r + \infty$, imperfect. (Fig. 41.) Hardness = 1.5, — 2.0. Sp. gr. = 2.2, — 2.4.

Description.—Most frequent colours white and grey; occurs also yellow, red, blue, green, brown, and even black. Occurs in regular crystals; in granular, scaly-granular, and fibrous distinct concretions; massive, disseminated, and dentiform. Lustre alternates from splendent to glimmering, and is pearly. Fracture splintery. Fragments indeterminate, angular, and blunt-edged. Alternates from transparent to translucent on the edges.

The transparent and highly crystallized varieties are named *selenite*; those in granular concretions, *foliated granular gypsum;* those disposed in fibrous concretions, *fibrous gypsum;* the splintery fracture characterizes the *compact gypsum;* while those varieties composed of scaly-granular concretions form the subspecies named *scaly-foliated gypsum.* Some varieties, composed of fine scaly or dusty and slightly cohering particles, are named *earthy gypsum.*

Constituent Parts.—Lime 33.0. Sulphuric acid 44.8. Water 21.0.—*Bucholz.*

Geognostic and Geographic Situations.—*Selenite,* the purest subspecies, occurs most frequently in what are called the gypsum and salt formations of the secondary and tertiary classes of rocks; also in metalliferous veins of different descriptions, and in various alluvial clays and marls. The *foliated granular* subspecies occurs in beds, in transition and secondary rocks,—being in the former intermixed with mica, in the latter with quartz, boracite, &c. The *compact variety* occurs in considerable abundance along with the granular in the secondary gypsum formation, and the *fibrous* is disposed in veins in the same formation, which also contains the *scaly-foliated* and the *earthy kinds.*

The salt mines in England afford examples of nearly all the subspecies, and several of them are also met with in Scotland.

2. *Prismatic Gypsum, or Anhydrite.*

Prismatisches Gyps-Haloide, *Mohs.*—Muriacit, *Werner.*—Chaux Anhydro-Sulphatée, *Haüy.*

Specific Character.—Prismatic. Pyramid = 121° 32′; 108° 35′; 99° 7′. Cleavage, $\breve{Pr} - \infty$. $\bar{Pr} + \infty$, very perfect. Less perfect $P - \infty$. Traces of $P + \infty$. (Fig. 29, 28, 27, 30.) Hardness = 3.0,—3.5. Sp. gr. = 2.7,—3.0.

Description.—Colours white, blue, red, and grey. Occurs crystallized; in granular, fibrous, and lamellar concretions; massive, and vermicularly convoluted or contorted. Lustre alternates from splendent to glistening, and is pearly. Fracture splintery and conchoidal. Alternates from transparent to translucent on the edges.

Constituent Parts.—Lime 41.75, sulphuric acid 55.0, muriate of soda, 1.0. *Klaproth.*

Geognostic and Geographic Situations.—Occurs massive, and in beds in the salt and secondary gypsum formations. Is frequently intermixed with rock-salt, also with stinkstone, saliniferous clay, and occasionally with ores of different kinds. Some varieties are met with in transition and primitive rocks. Several of the varieties are found in the red sandstones of England and Scotland.

Genus II.—Cryolite.

Prismatic. Cleavage in three directions, perpendicular to each other, of which one is more perfect than the others. Hardness = 2.5, — 3.0. Sp. gr. = 2.9, — 3.0.

1. *Prismatic Cryolite.*

Kryolite, *Werner.*—Prismatisches Kryon-Haloide, *Mohs.* —Alumine fluatée alcaline, *Haüy.*

Specific Character.—Prismatic. Pyramid unknown. Cleavage, $P - \infty$. Less distinct, $\breve{Pr} + \infty$. $\bar{Pr} + \infty$. Traces of P. (Fig. 27, 29, 28.)

Description.—Colours white, brown, and red. Occurs massive, disseminated, and in lamellar concretions. Internally shining, and lustre vitreous, inclining to pearly. Fracture uneven. Fragments cubical. Translucent. Brittle, and easily frangible.

Constituent Parts.—Alumina 24.40, soda 31.35, fluoric acid 44.25. *Berzelius.*

Geognostic and Geographic Situations. It has hitherto been found only in West Greenland, where it occurs in gneiss, associated with iron-pyrites and galena or lead-glance.

Genus III.—Alumstone.

Alaun-Haloide, *Mohs.*—Alaunstein, *Werner.*

Rhombohedral. Hardness = 5.0. Sp. gr. = 2.4,—2.6.

1. *Rhombohedral Alumstone.*

Rhomboidrisches Alaun-Haloide, *Mohs.*

Specific Character.—Rhombohedral. Rhomboid unknown. Cleavage R—∞. R.

Description.—Colours white, red, and rarely grey. Occurs massive, porous, and vesicular. Lustre feebly glimmering. Fracture uneven. Feebly translucent on the edges. Brittle, and easily frangible.

Constituent Parts.—Mont d'Or in Auvergne. Sulphuric acid 27.05, alumina 31.80, potash 5.79, silica 28.40, water and loss 3.72. *Cordier.*

Geognostic and Geographic Situations.—Occurs in beds and large irregular masses in porphyry in Hungary, and in veins and drusy cavities in aluminous rocks at Tolfa, near to Civita Vecchia.

Genus IV.—Fluor.

Tessular. Hardness = 4.0. Sp. gr. = 3.0,—3.3.

1. *Octahedral Fluor.*

Octaedrisches Flus-Haloide, *Mohs.*—Chaux fluatée, *Haüy.*

Specific Character.—Tessular. Cleavage octahedral.

Description.—Colours white, grey, black, blue, green, yellow, red, and brown. Occurs regularly crystallized, in granular, prismatic, and lamellar concretions, massive, and disseminated. Lustre from splendent to feebly glimmering, and vitreous. Fracture even, inclining to splintery and to conchoidal. More or less transparent and translucent. Brittle, and easily frangible.

The varieties, with even fracture and feeble lustre, are named *compact fluor;* those in which the cleavage is distinct are named common or *foliated fluor;* and some rare, dull, earthy, and loosely aggregated varieties, which occur, incrusting other minerals, are described under the name *earthy fluor.*

Constituent Parts.—Lime 67.75, fluoric acid 32.25. *Klaproth.*

Geognostic and Geographic Situations.—It occurs in veins and beds in primitive gneiss, mica-slate, and clay-slate, in various remarkable metalliferous formations of cobalt, silver, tin, lead, copper, &c.; less frequently in transition rocks, very abundantly in some secondary rocks, as limestone, and rarely in secondary porphyries. It is a rare mineral in Scotland, its principal localities being Monaltree, in Aberdeenshire, Banffshire, Papa Stour in Shetland, and Gourock in Renfrewshire. It is very abundant in several of the mining districts in England.

Genus V.—Apatite.

Rhombohedral. Hardness = 5.0. Sp. gr. = 3.0,—3.3.

1. *Rhombohedral Apatite.*

Rhomboedrishes Flus-Haloide, *Mohs.*—Apatit, *Werner.*—Chaux phosphatée, *Haüy.*

Specific Character.—Rhombohedral. R = 88° 41′.

Combination, dirhombohedral. 2. R = 131° 14′; 111° 20′, $(P + n)^m$ hemi-dirhombohedral, with parallel planes. Cleavage, R — ∞ . $P_{,}$ + ∞ . H = 5. Sp. gr. = 3.0, — 3.3.

Description.—Colours white, green, blue, red, yellow, and brown. Occurs regularly crystallized; in concretions which are granular, lamellar, and fibrous; massive and disseminated; lustre resinous, and varying from splendent to glimmering. Fracture conchoidal and uneven. Alternates from transparent to feebly translucent on the edges. Brittle, and easily frangible.

One set of varieties, in which the cleavage is very distinct, is named *foliated apatite;* another, in which the fracture is conchoidal, is named *conchoidal apatite;* and the varieties in which the fracture is uneven are named *phosphorite.*

Constituent Parts.—Lime 55.75, phosphoric acid 44.25. *Klaproth.*

Geognostic and Geographic Situations.—Occurs in gneiss, near Kincardine, in Ross-shire; in the same rock in the Shetland Islands; and in veins in greenstone, in the Island of Rume. Several varieties are met with in Cornwall.

Genus VI.—Limestone.

Kalk-Haloide, *Mohs.*

Rhombohedral, prismatic. Cleavage, rhombohedro-paratomous, prismatoidal. Hardness = 3.0, — 4.5. Sp. gr. = 2.5, — 3.2. If prismatic, the specific gravity = 3.0 and less. If the hardness is above 4.0, the specific gravity = 2.8 and more. If specific gravity = 2.9 and more, the hardness = 2.8 and more.

1. *Aragonite. Prismatic Limestone.*

Prismatisches Kalk-Haloide, *Mohs.*—Aragon, *Werner.* —Aragonite, *Haüy.*

Specific Character.—Prismatic. Pyramid = 113° 44′; 93° 43′; 122° 10′. P + ∞ = 105° 23′. Cleavage, $\bar{Pr} - 1$ = 109° 28′. $(\breve{Pr} + \infty)^3$ = 64° 4′. More distinct, $\bar{Pr} + \infty$. (Fig. 42.) Hardness = 3.5, — 4.0. Sp. gr. = 2.6, — 3.0.

Description.—Colours white, grey, green, and violet-blue. Occurs regularly crystallized; also in prismatic concretions and massive. Lustre vitreous, inclining to resinous, and shining and glistening. Fracture conchoidal passing into uneven. Translucent and transparent. Brittle, and easily frangible.

Constituent Parts.—Carbonate of lime 94.82, carbonate of strontites 4.08, water 0.98. *Stromeyer.*

Geognostic and Geographic Situations.—Occurs along with galena in the lead-mines of Leadhills, and in secondary trap-rocks in different parts of Scotland.

2. *Calcareous Spar. Rhombohedral Limestone.*

Rhomboedrisches Kalk-Haloide, *Mohs.*—Chaux carbonatée, *Haüy.*—Kalk-spath, *Werner.*

Specific Character.—Rhombohedral. Rhombohedron = 105° 5′. Cleavage, R. Hardness = 3.0. Sp. gr. = 2.5, — 2.8.

Constituent Parts.—The purest kind of this species, named calcareous spar, is composed of lime 56.15, carbonic acid 43.70.

Description.—Colours very varied, but of all the tints, white and grey are the most frequent; besides these, the following kinds also occur, viz. red, blue, green, yellow, brown, and rarely black. Occurs regularly crystallized, and of all known minerals, exhibits the greatest number of varieties of the rhombohedral series of crystallization; also in granular, prismatic, tabular, and globular distinct concretions. Its other forms are massive, disseminated, globular, botryoidal, reniform, tuberose, stalactitic, tabular, claviform, coralloidal, cellular, and curtain-shaped. Its lustre varies from splendent to dull, and is vitreous, inclining sometimes to pearly, sometimes to resinous. Fracture splintery, conchoidal, earthy, and uneven. Alternates from transparent to opaque. Generally brittle, and easily frangible.

Those varieties which are regularly crystallized, and possess high degrees of transparency, are named *calcareous spar;* those in angulo-granular distinct concretions, with a lower lustre and transparency than the former, are the *foliated granular limestone*, or *crystalline marble* of authors; the varieties with splintery or conchoidal fractures are named *compact limestone;* other varieties having a grey or brown colour, dull earthy fracture, and which, on rubbing, give out a sulphureo-bituminous smell, are named *stinkstone;* the black varieties in granular and prismatic concretions, or with a compact fracture, with a glimmering or shining lustre, and low degree of translucency on the edges or complete opacity, and which, on rubbing, yield a sulphureo-bituminous odour, are named *anthraconite;* those limestones which have the oolitic structure, or are composed of spherical granular concretions, set on a marly basis, are named *roestone* or *oolite;* the soft varieties with earthy fracture and white colour are named *chalk;* the varieties with dull fracture surface, in loosely cohering pieces and crusts, and so light as nearly to swim upon water, are named *agaric mineral;* the fibrous varieties are named *common fibrous limestone* or *satin spar*, and *fibrous calc-sinter*, or calcareous alabaster; the earthy-looking opaque varieties that occur in many particular external shapes, near calcareous springs, and on the borders of lakes, are named *calc-tuff*; while the varieties in spherical round granular concretions, and concentric lamellar concretions, found near hot springs, are named *peastone;* the *slate-spar* of mineralogists is a variety in lamellar concretions, with a beautiful pearly lustre and feeble translucency; varieties with earthy fracture, more or less inclined to splintery and conchoidal, are named *marl;* and, lastly, the opaque dull black varieties, with slaty fracture, are named *bituminous marl-slate.*

Geognostic Situation.—This mineral is one of the most abundant and widely distributed with which we are acquainted. Calcareous spar, one of its principal kinds, occurs in every rock from granite to the newest member of the secondary series. It generally occurs in veins with numerous metalliferous minerals, and assists in an eminent degree in characterizing the vast host of mineral veins in primitive, transition, and secondary rocks. In the state of granular, foliated, and compact limestones, also in the form of oolite, chalk, and lucullite, it forms beds, hills, mountains, and even ranges of mountains. The beautiful fibrous limestone or satin-spar, occurs in veins in clay-slate, and in rocks of the coal formation, while the varied forms of calc-sinter are met with ornamenting the walls and floors of caverns in limestone and other formations; calc-tuff abounds around cold, and also warm springs, frequently incrusting organic bodies, forming the calcareous incrustations so well known to mineralogists. Peastone is also a production of warm springs. The white, porous, and nearly supernatent mineral agaric incrusts rocky cliffs, particularly in limestone hills, and the rare variety, named slate-spar, has hitherto been found only in primitive limestone.

Geographic Situation.—England and Scotland abound in interesting varieties of calcareous-spar, and the mountains, hills, and valleys of Great Britain afford numerous localities of many of the different kinds of marble, limestone, chalk, marl, Lucullite, and oolite, while its calcareous springs, and caves and caverns, exhibit numerous deposits of calc-tuff and of calc-sinter.

3. *Dolomite. Macrotypous* [1] *Limestone.*

Macrotypes Kalk-Haloide, *Mohs.*—Braunspath, Rhombspath, Dolomite, *Werner,*—Chaux carbonatée ferrifère perlée, Chaux carbonatée magnesifère, *Haüy.*

Specific Character.—Rhombohedral. Rhombohedron 106° 15′. Cleavage, rhombohedral. Hardness = 3.5,—4.0. Sp. gr. = 2.8,—2.95.

Description—Colours white, grey, brown, red, and green. Occurs crystallized in rhomboids, in granular and prismatic concretions; massive, disseminated, globular, stalactitic, reniform, and with tabular and pyramidal impressions. Lustre varies from shining to glimmering, and is pearly, sometimes inclining to vitreous. Fracture splintery, conchoidal, and slaty. Varies from transparent to translucent on the edges. Brittle, and easily frangible.

The white varieties in small and fine granular concretions, which are sometimes so loosely aggregated as to separate by the mere pressure of the finger, are the *dolomite-marble* of mineralogists; the *magnesian limestone* of

[1] From μακρὸς *long;* and τύπος, *the type,* (fundamental form).

England is a dolomite with yellow or yellowish brown colours; the green varieties are described under the name *Miemite*, from Miemo in Tuscany, where they were first found; the brown, red, reddish-white, and pearl-grey varieties, with very distinct pearly lustre, are arranged together, and described under the names *brown spar* and *pearl spar*.

Constituent Parts.—Carbonate of lime 68.0, carbonate of magnesia 25.5, alumina 2.0, carbonate of iron 1.0, water 2.0. *Klaproth.*

Geognostic and Geographic Situations.—The dolomite marble occurs in the island of Iona; the yellow dolomite constitutes a secondary limestone formation very abundant in England; the brown spar and pearl spar are not unfrequent in the lead mines of Scotland and England.

4. *Breunnerite. Brachytypous*[1] *Limestone.*

Brachytypes Kalk-Haloid, *Mohs.*—Rautenspath, *Werner.* —Chaux carbonatée magnesifere, *Haüy.*

Specific Character.—Rhombohedral. Rhombohedron = 107° 22′. Cleavage rhombohedral. Hardness = 4.0, —4.5. Sp. gr. = 3.0,—3.2.

Description.—Colours white, grey, and yellow. Occurs crystallized in rhombs; also massive and disseminated. Lustre splendent and vitreo-pearly. Fracture imperfect conchoidal. More or less translucent. Brittle, and easily frangible.

Constituent Parts—Magnesia, 41.06, protoxide of iron 8.57, oxide of manganese 0.43, carbonic acid 48.94. *Stromeyer.*

Geognostic and Geographic Situations.—Occurs imbed-

[7] From βραχος, *short*; and τύπος, *the type.*

ded in chlorite-slate on the banks of Loch Lomond, and associated with galena, copper-pyrites, and blende, near Newton Stewart, in Galloway. It is named in honour of Count Breunner.

5. *Ankerite. Paratomous Limestone.*

Specific Character.—Rhombohedral. Cleavage parallel to a rhombohedron of 106° 12′. H = 3.5,—4.0. Sp. gr. = 2.95,—3.1. Is a compound of carbonate of lime and carbonate of iron.

Geognostic and Geographic Situations.—Found in the Styrian mines. Is named in honour of Professor Anker.

Order II.—BARYTE.

Not metallic. No metallic-pearly lustre. Streak white, grey, or orange-yellow. H = 2.5,—5.0. Sp. gr. = 3.3, —7.3. If cleavage monotomous, the specific gravity = 4.0 and less, or = 5.0 and more. Not elastic, not flexible, in thin folia. If lustre adamantine or imperfect metallic, the sp. gr. = 5.0 and more. If streak orange-yellow, the sp. gr. = 6.0 and more. If H = 5.0, and sp. gr. less than 4.5, the streak is white. If the sp. gr. is less than 4.0, and H = 5.0, the cleavage is diprismatic.

Genus I.—Sparry Iron.

Rhombohedral. Cleavage paratomous. Hardness =3.5, —4.5. Sp. gr.=3.3,—3.9.

1. *Rhombohedral Sparry Iron.*

Brachytyper Parachros Baryte, *Mohs.*—Spath Eisenstein, *Werner.*—Fer Oxydé Carbonaté, *Haüy.*

Specific Character.—Rhombohedral. Rhombohedron

= 107° 0′. Cleavage rhombohedral. H = 3.5,—4.5. Sp. gr. = 3.3,—3.9.

Description.—Colours yellow, white, brown, and black. Occurs crystallized in rhombohedrons; also in granular concretions, massive and disseminated. Internally, lustre pearly, and varying from shining to glimmering, and even to splendent. Fracture sometimes splintery. Translucent on the edges. Rather brittle, and easily frangible.

Constituent Parts.—Protoxide of iron 57.50, carbonic acid 36.00, oxide of manganese 3.30, lime 1.25. *Klaproth.*

Geognostic and Geographic Situations.—Occurs in metalliferous veins, and in common veins, in primitive, transition, and secondary rocks in different parts of Great Britain and Ireland. The common *clay ironstone* of the coal formation, the most important of ores, is an impure kind, of this species.

Genus II.—Red Manganese.

Hardness = 3.5. Sp. gr. = 3.3,—3.6.

1. *Rhombohedral Red Manganese.*

Macrotyper Parachros Baryte, *Mohs.*—Rother Braunstein, *Werner.* Manganese Oxidé Carbonaté, *Haüy.*

Specific Character.—Rhombohedron = 106° 51′. Cleavage, rhombohedral.

Description.—Colours red and brown. Occurs in granular concretions, also in fibrous concretions, which are scopiformly and stellularly arranged, massive, and reniform. Lustre varies from shining to glimmering, and pearly. Fracture splintery. More or less translucent on the edges;

in some rare varieties translucent. Brittle, and rather easily frangible.

The varieties with distinct cleavage are named *foliated red manganese*; those in fibrous concretions, *fibrous red manganese*; and the splintery varieties, *compact red manganese.*

Constituent Parts.—Oxide of manganese 54.60, carbonic acid 33.75, oxide of iron 1.87, silica 4.37, lime 2.50. *Du Menil.*

Geognostic and Geographic Situations.—Occurs at Kapnic in Transylvania, and at Catharinenberg in Siberia.

Genus III.—Calamine.

Zinc-Baryt, *Mohs.*

Rhombohedral. Prismatic. Hardness = 5.0. Sp. gr. =3.3,—4.5. If rhombohedral, the sp. gr. above 4.0.

1. *Prismatic Calamine, or Silicate of Zinc.*

Prismatischer Zinc-Baryt, *Mohs.* Zink Oxydé; *Haüy.*

Specific Character.—Prismatic. Pyramid = 134° 59′; 99° 56′; 96° 56′. $P + \infty$ 118° 29′. Cleavage, $\breve{P}r$ = 120°. More distinct, $(\breve{P}r + \infty)^3$ = 80° 4′. (Fig. 43.) Hardness 5.0. Sp. gr. =3.3,—3.6.

Description.—Most frequent colours white and yellow; also green, grey, yellow, and brown; and with curved striped colour delineations. Occurs regularly crystallized, and in distinct concretions, which are scopiform radiated, and scopiform fibrous, granular, and curved lamellar. Massive, in crusts, stalactitic, reniform, botryoidal, and cellular. Internally alternates from glistening to dull, and lustre pearly, inclining to adamantine. Fracture small and fine-grained, uneven. Varies from transparent to opaque.

Constituent Parts.—Oxide of zinc 66.83, silica 24.89, water 7.46. *Berthier.*

Geognostic and Geographic Situations.—Occurs in veins of galena, in greywacke, and clay-slate, and in beds, and imbedded masses in secondary limestone. Is found in the lead mines of Wanlockhead, and in the mines of Flintshire and Leicestershire.

2. *Rhombohedral Calamine or Carbonate of Zinc.*

Rhomboedrischer Zinc-baryt, *Mohs.*—Galmei, *Werner.*—Zinc carbonaté, *Haüy.*

Specific Character.—Rhombohedral. Rhombohedron = 110° (nearly.) Cleavage rhombohedral. Hardness = 5.0. Sp. gr. = 4.2,—4.5.

Description.—Colours white, grey, green, yellow, and brown. Occurs regularly crystallized, and in distinct concretions, which are radiated, granular, and curved lamellar; massive, corroded, reniform, stalactitic, botryoidal, and cellular. Internally ranges from shining to dull, and is pearly. Fracture uneven, splintery, and flat conchoidal. Ranges from transparent to opaque.

Constituent Parts.—Oxide of zinc 65.20, carbonic acid 34.80. *Smithson.*

Geognostic and Geographic Situations.—Occurs in beds, veins, nests, filling up or lining hollows, in transition limestone and in secondary limestone. Derbyshire, Somersetshire, Flintshire, and Durham, afford numerous localities of this mineral.

Genus IV.—Tungsten, or Scheelium.

Pyramidal. Hardness = 4.0,—4.5. Sp. gr. = 6.0,—6.1.

1. *Pyramidal Tungsten.*

Pyramidaler Scheel-Baryt, *Mohs.*—Schwerstein, *Werner.* —Scheelin Calcaire, *Haüy.*

Specific Character.—Pyramidal. Pyramid $= 107° 27'$; $113° 35'$. Combination hemi-pyramidal with parallel planes. Cleavage, $P + 1 = 100° 8'$; $130° 20'$. P. Traces in direction of $P - \infty$.

Description.—White is the principal colour; but other varieties, as brown and orange yellow, occasionally occur. Sometimes crystallized, and also in distinct concretions, which are granular, prismatic, and curved lamellar. Occurs massive and disseminated. External lustre shining and splendent; internal lustre shining and resinous. Fracture uneven or conchoidal. More or less translucent, seldom transparent.

Constituent Parts.—Lime 19.40, oxide of scheelium 80.24. *Berzelius.*

Geognostic and Geographic Situations.—Occurs along with tinstone, wolfram, magnetic iron-ore, and brown iron-ore, in Cornwall, &c.

Genus V. Baryte.

Prismatic. Hardness $= 3.0, - 4.0$. Sp. gr. $3.6, - 4.7$.

1. *Strontianite, or Carbonate of Strontites. Peritomous Baryte.*

Peritomer Hal-Baryt, *Mohs.*—Strontian, *Werner.*—Strontiane Carbonaté, *Haüy.*

Specific Character.—Prismatic. Pyramid unknown. Cleavage, $P + \infty = 117° 19'$; less distinct $\mathring{P}r$. Traces according to $\mathring{P}r + \infty$. (Fig. 46.) Hardness $= 3.5$. Sp. gr. $3.6, - 3.8$.

Description.—Colours green and grey. Occurs regularly crystallized, and in distinct concretions, which are scopiform radiated, and scopiform fibrous. Lustre shining, glistening, and pearly. Fracture uneven. More or less translucent. Brittle, and easily frangible.

Constituent Parts.—Strontia 61.21, carbonic acid 30.20, water 8.59. *Hope.*

Geognostic and Geographic Situations.—Occurs at Strontian in Scotland in veins of lead-glance that traverse gneiss.

2. *Witherite, or Carbonate of Barytes. Di-prismatic Baryte.*

Di-prismatischer Hal-Paryt, *Mohs.*—Witherit, *Werner.* —Baryte carbonatée, *Haüy.*

Specific Character.—Prismatic. Pyramid unknown. Cleavage, $P.+\infty=118^\circ\ 30'$. $\overset{\smile}{Pr}+\infty$. $\overset{\circ}{Pr}+1$. (Fig. 46.) Hardness $=3.0,$—3.5. Sp. gr. $=4.2,$—4.4.

Description.—Colours white, grey, and yellow. Occurs regularly crystallized, and also in distinct concretions, which are radiated and granular. Occurs massive, cellular, globular, botryoidal, reniform, and stalactitic. Lustre shining and resinous. Fracture uneven, inclining to splintery. Translucent. Brittle, and easily frangible.

Constituent Parts.—*Witherit.* Baryta 79.66, carbonic acid 20.00, water 0.33. *Bucholz.*

Geognostic and Geographic Situations.—Occurs abundantly in the lead mines of Cumberland, and at Anglesark in Lancashire.

3. *Heavy-Spar, or Sulphate of Barytes. Prismatic Heavy Spar.*

Prismatischer Hal-baryte, *Mohs*—Schwerspath, *Werner.* —Baryte sulphatée, *Haüy.*

Specific Character.—Prismatic. $\breve{P}r = 105°6'$; $(\breve{P}r + \infty)^2 = 77° 27'$. Cleavage $\bar{P}r = 78° 18'$. $\breve{P}r + \infty$. (Fig. 45.) Less distinct, $P - \infty$. (Fig. 33, 20, 27, 28.) Hardness=3.0, —3.5. Sp. gr.=4.1,—4.7.

Description.—Colours white, grey, black, blue, green, yellow, red, and brown. Occurs regularly crystallized; also in granular, lamellar, fibrous, and prismatic distinct concretions; massive, disseminated, reniform, botryoidal, and globular. Lustre alternates from splendent to glimmering, and resinous or pearly, inclining to vitreous. Fracture uneven, splintery, and earthy. Alternates from transparent to opaque. Brittle, and very easily frangible.

The varieties with uneven and splintery fracture are named *compact h. spar;* those in fine granular concretion *granular h. spar;* the lamellar varieties are named *straight or curved lamellar h. spar,* according to the direction of the lamellar concretion; the fibrous varieties *fibrous h. spar;* the radiated varieties *radiated h. spar;* those in prismatic concretions *prismatic h. spar;* the varieties which, on rubbing, emit a hepatic smell, *fetid h. spar, or hepatite;* and those which occur in the earthy or powdery state, *earthy h. spar.*

Constituent Parts.—Baryta 66.00, sulphuric acid 34.00. *Berthier.*

Geognostic and Geographic Situations.—It occurs in veins, either alone or associated with various metalliferous formations of silver, copper, lead, cobalt, antimony, manganese, zinc, arsenic, iron, &c. in rocks of the primitive, transition, and secondary classes. There are numerous localities of this mineral in Scotland, England, and Ireland.

4. *Celestine, or Sulphate of Strontites. Prismatoidal Baryte.*

Prismatoidischer Hal-baryt, *Mohs.*—Coelestin, *Werner.* —Strontiane sulphatée, *Haüy.*

Specific Character. — Prismatic. $\breve{P}r = 103° \ 58'$. $(\breve{P} + \infty .)^2 = 78° \ 35'$. Cleavage, $\bar{P}r = 76° \ 2'$. More distinct. $\breve{P}r + \infty$. Less distinct P—∞. Hardness =3.0, —3.5. Sp. gr. =3.6,—4.0.

Description.—Colours white, blue, and red. Occurs regularly crystallized; in granular, fibrous, and radiated distinct concretions; massive and stalactitic. Lustre alternates from splendent to glimmering, and is pearly. Ranges from transparent to translucent on the edges.

Constituent Parts.—Strontia 58.00, sulphuric acid 42.00. *Klaproth.*

Geognostic and Geographic Situations.—Occurs in limestone, in red sandstone, and gypsum, where it is associated with sulphur, and in vesicular cavities in amygdaloid. It has been found in sandstone near Inverness; in the amygdaloid of the Calton Hill; in the neighbourhood of Bristol; and near Knaresborough, in Yorkshire.

Genus VI.—Lead-Spar.

Blei-baryt, *Mohs.*

Rhombohedral, pyramidal, prismatic. Hardness = 2.5, —4.0. Sp. gr. =6.0,—7.3. If the hardness is above 3.5, the sp. gr. =6.5, and more.

1. *White Lead-Spar, or Carbonate of Lead. Di-prismatic Lead-Spar.*

Di-prismatischer Blei-baryt, *Mohs.*—Plomb carbonatée, *Haüy.*

Specific Character.—Prismatic. P = 130° 0′; 108° 28′; 92° 19′. Cleavage, $\breve{P}r$ = 117° 13′. $(\breve{P}r + \infty)^3$ = 69°20′. (Fig. 43.) Hardness = 3.0,—3.5. Sp. gr. = 6.3,—6.6.

Description.—Principal colour white, occurs also black, yellow, brown, and grey. Occurs regularly crystallized; massive and cellular. Lustre externally ranges from splendent to shining; internally from shining to glistening, and is adamantine, inclining more or less to resinous and imperfect metallic. Fracture uneven and conchoidal. Ranges from transparent to opaque. Streak greyish-white.

Constituent Parts.—Oxide of lead 82.0, carbonic acid 16.0, water 2.0. *Klaproth.*

Geognostic and Geographic Situations.—Occurs in veins in granite, gneiss, mica-slate, and clay-slate; also in limestone, greywacke, and in various secondary formations. Well known localities are Leadhills and Wanlockhead.

2. *Green and Brown Lead-Spar, or Phosphate of Lead. Rhombohedral Lead-Spar.*

Rhomboedrischer Blei-Baryt, *Mohs.*—Plomb phosphaté, *Haüy.*

Specific Character.—Rhombohedral. R = 88° 29′; 2 (R) = 131° 5′; 111° 48′. P = 142° 12′; 80° 44′. Cleavage, P. P + ∞. Both imperfect. Hardness = 3.5,—4.0. Sp. gr. = 6.9,—7.3.

Description.—Colours green, brown, and sometimes yellow and white. Occurs regularly crystallized; also in granular, radiated, and curved lamellar concretions. Lustre externally shining, internally glistening, and resinous. Fracture uneven or splintery. Ranges from translucent to translucent on the edges. Streak white.

Constituent Parts.—Oxide of lead 78.5, phosphoric acid 19.73, muriatic acid 1.65.

Geognostic and Geographic Situations.—Occurs in veins, and most abundantly in their upper part, in various rocks of the primitive, transition, and secondary classes, where it is associated with galena, white lead-spar, &c. Leadhills and Wanlockhead are well known localities.

3. *Red Lead-Spar, or Chromate of Lead. Hemi-Prismatic Lead-Spar.*

Hemi-prismatischer Blei-Baryt, *Mohs.*—Roth Bleierz, *Werner*—Plomb chromaté, *Haüy.*

Specific Character.—Prismatic. Pyramid unknown. Combination hemi-prismatic. Cleavage, $P + \infty = 90^\circ$ (nearly). $\breve{P}r + \infty$, $\bar{P}r + \infty$. (Fig. 30, 29, 28.) Hardness $= 2.5$. Sp. gr. $= 6.0, - 6.1$.

Description.—Colour hyacinth-red. Occurs regularly crystallized; also massive, and in flakes. Internally shining or splendent, and lustre adamantine. Fracture uneven, sometimes imperfect conchoidal. More or less translucent. Streak between lemon-yellow and orange-yellow. Nearly sectile, and easily frangible.

Constituent Parts.—Oxide of lead 63.96, chromic acid 36.40. *Thenard.*

Geognostic and Geographic Situations.—Occurs in a granular quartzy rock, associated with brown ironstone, iron-pyrites, green lead-spar, native gold, galena, and quartz, in Siberia; also in sandstone in the Brazils.

4. *Yellow Lead-Spar, or Molybdate of Lead. Pyramidal Lead-Spar.*

Pyramidaler Blei-Baryt, *Mohs.*—Gelb Bleierz, *Werner.*—Plomb molybdaté, *Haüy.*

Specific Character.—Pyramidal. Pyramid $= 99^\circ\ 40'$;

131° 35′. Cleavage P. less distinct. P—∞. Hardness =3.0. Sp. gr. =6.5,—6.9.

Description.—Colour yellow. Occurs crystallized; also massive, in crusts, and cellular. Externally generally splendent and shining; internally shining or glistening, and lustre resino-adamantine. Fracture uneven and conchoidal. Translucent, or translucent on the edges.

Constituent Parts.—Oxide of lead 58.40, molybdic acid 38.00, oxide of iron 2.08. *Hatchet.*

Geognostic and Geographic Situations.—Occurs in compact limestone, at Bleiberg in Carinthia; also in France and Saxony.

5. *Sulphate of Lead, or Prismatic Lead-Spar.*

Prismatischer Blei-Baryt, *Mohs.*—Vitriol Bleierz, *Werner.*—Plomb sulphaté, *Haüy.*

Specific Character.—Prismatic. Pyramid = 104° 55′; $(\breve{P}+\infty)^2$ = 78° 45′. Cleavage, $\bar{P}r$ = 76° 11′. More distinct $\breve{P}r+\infty$. (Fig. 45.) Hardness = 3.0. Sp. gr. 6.2, —6.3.

Description.—Colour white, seldom green or wine yellow, or blue (owing to blue malachite). Occurs regularly crystallized; in granular distinct concretions; also massive and disseminated. Lustre adamantine, and ranging from splendent to shining. Fracture conchoidal. Transparent or translucent.

Constituent Parts.—Oxide of lead 72.47, sulphuric acid 26.09, water 0.12, silica 0.51. *Stromeyer.*

Geognostic and Geographic Situations.—Occurs in galena veins, at Leadhills and Wanlockhead; and also in the Pary's Mine in Anglesea.

Order III.—KERATE.[1]

Not metallic. Streak white or grey. No single distinct cleavage. Hardness = 1.0,—2.0. Sp. gr. = 5.5,—6.5.

Genus I.—Corneous Silver, or Chloride of Silver.

Hardness = 1.0,—2.0. Sp. gr. = 5.5,—6.5.

1. *Hexahedral Corneous Silver, or Muriate of Silver.*

Hexedrisches Perl Kerat, *Mohs.*—Hornerz, *Werner.*—Argent muriaté, *Haüy.*

Specific Character.—Tessular. Cleavage not visible. Malleable. Sectile.

Description.—Colour pearl-grey, which passes into white, blue, and green; on exposure to light becomes brown. Occurs crystallized; in prismatic and granular concretions; massive, and in flakes. Lustre ranges from shining to glistening and is resinous. Fracture conchoidal. Translucent, or feebly translucent on the edges. Becomes shining in the streak.

Constituent Parts.—Silver 76.0, chlorine 24.0.

Geognostic and Geographic Situations.—Occurs in silver-mines in Siberia, America, and has also been found in Cornwall.

Genus II.—Corneous Mercury, or Muriate of Mercury.

Hardness = 1.0,—2.0. Sp. gr. 6.4,—6.5.

[1] From κερας, horn, as the minerals of this order resemble horn in external aspect.

1. *Pyramidal Corneous Mercury.*

Pyramidales Perl Kerat, *Mohs.*—Quecksilber Hornerz, *Werner.*—Mercure muriaté, *Haüy.*

Specific Character.—Pyramidal. P = 98° 4′; 136° 0′. Cleavage P + ∞, imperfect. Sectile.

Description.—Colour grey. Occurs generally in very minute crystals in vesicular cavities. Lustre shining and adamantine. Faintly translucent.

Constituent Parts.—Oxide of mercury 76.0, muriatic acid 16.4, sulphuric acid 7.6. *Klaproth.*

Geognostic and Geographic Situations.—Occurs in the mercury-mines of Almaden, Idria, and Bohemia.

Order IV.—MALACHITE.

Not metallic. Colour blue, green, brown. No single distinct faces of cleavage. Hardness = 2.0,— 5.0. Sp. gr. = 2.0,—4.6. If brown, in colour or in streak, the hardness = 3.0, and less; and the specific gravity above 2.5. If white in the streak, the specific gravity = 2.2, and less; and the hardness under 3.0.

Genus I.—Copper Green, or Chrysocolla.

Staphylin Malachit, *Mohs.*

Uncleavable. Hardness = 2.0,— 3.0. Sp. gr. = 2.0, —2.2.

1. *Uncleavable Copper-Green.*

Untheilbarer Staphylin-Malachite, *Mohs.*—Kupfergrün, *Werner.*—Cuivre carbonaté, *Haüy.*

Specific Character.—Reniform, botryoidal, massive. Fracture conchoidal, shining. Streak white.

Description.—Colour green. Occurs massive, disseminated, in crusts, reniform, and botryoidal. Lustre shining, glistening, and resinous. Fracture small, conchoidal. Ranges from translucent to translucent on the edges. Colour does not change in the streak.

Constituent Parts.—Oxide of copper 50.0, silica 26.0, water 17.0, carbonic acid 7.0. *Klaproth.*

Geognostic and Geographic Situations.—Occurs in copper-mines in Cornwall.

Genus II.—Liriconite.[1]

Lirikon-Malachite, *Mohs.*

Tessular, prismatic. Hardness = 2.0,—2.5. Sp. gr. = 2.8,—3.0.

1. *Prismatic Liriconite, or Lenticular Arseniate of Copper.*

Prismatischer Lirikon-Malachite, *Mohs.* — Linsenerz, *Werner.*—Cuivre arseniaté, *Haüy.*

Specific Character.—Prismatic. $\bar{P}r = 71° 59'$. $P + \infty = 119° 45'$. Cleavage, $\bar{Pr}$. $P + \infty$, imperfect. (Fig. 35, 30.) Streak pale verdigris-green,...Sky-blue. Hardness = 2.0,—2.5. Sp. gr. = 2.8,—3.0.

Description.—Colour blue, inclining more or less to verdigris-green. Occurs regularly crystallized. Lustre shining, glistening, and pearly, passing to vitreous. Fracture uneven. Translucent. Brittle, and uncommonly easily frangible.

Geognostic and Geographic Situations.—Occurs in the copper-mines of Cornwall.

[1] From λειρος, pale, and κονια, dust (the streak).

2. *Hexahedral Liriconite, or Cubical Arseniate of Iron.*

Hexaedrischer Liricon-Malachite, *Mohs.* — Wurfelerz, *Werner.*—Fer arseniaté, *Haüy.*

Specific Character.—Tessular. Combination semi-tessular, with inclined planes. Cleavage, hexahedral. Streak pale olive-green,...brown. Hardness = 2.5. Sp. gr. = 2.9,—3.0.

Description.—Colour green. Occurs regularly crystallized, and massive. Lustre glistening and vitreo-resinous. Translucent, or translucent on the edges.

Constituent Parts.—Oxide of iron 40.56, arsenic acid 38.00, oxide of copper 0.60, silica 0.35, water 19.57. *Berzelius.*

Geognostic and Geographic Situations.—Occurs in the copper-mines of Cornwall.

Genus III.—Olivenite.

Oliven-Malachit, *Mohs.*

Prismatic. Cleavage very imperfect. Colour neither blue nor bright green. Streak olive-green—brown. Hardness = 3.0,—4.0. Sp. gr. = 3.6,—4.6.

1. *Prismatic Olivenite,* or *Prismatic Arseniate of Copper.*

Prismatischer Oliven-Malachit, *Mohs.*—Olivenerz, *Werner.*—Cuivre arseniaté, *Haüy.*

Specific Character.—Prismatic. $\breve{Pr} = 110° 50'$. $P + \infty = 92° 30'$. Cleavage, $\breve{Pr}$, $P + \infty$. (Fig. 30.) Streak olive-green,—brown. Hardness = 3.0. Sp. gr. = 4.2,—4.6.

Description.—Principal colour olive-green, also yellow, brown, and white. Occurs regularly crystallized; in concretions which are scopiform fibrous, angulo-granular, and curved lamellar; massive and in drusy crusts. Lustre ranges from splendent to glimmering, and is resinous, inclining to pearly, or pearly. Ranges from transparent to opaque.

Constituent Parts.—Oxide of copper 50.62, arsenic acid 45.00, water 3.50. *Klaproth.*

Geognostic and Geographic Situations.—Occurs in the copper-mines of Cornwall.

2. *Diprismatic Olivenite or Phosphate of Copper of Libethen.*

Diprismatischer Oliven-Malachit, *Mohs.*

Specific Character.—Prismatic. $\breve{P}r = 111° 58'$. $P + \infty = 95° 2'$. Cleavage $+ \breve{P}r + \infty$. $\bar{P}r + \infty$, also $\breve{P}r$ and $P + \infty$. Streak olive-green. Hardness = 4.0. Sp. gr. = 3.6,—3.8.

Description.—Colours grass, olive, leek, and pistachio-green. Occurs regularly crystallized. Lustre vitreous and pearly. Fracture conchoidal. Ranges from semitransparent to translucent.

Constituent Parts.—Oxide of copper 63.0, phosphoric acid 28.7, water 8.4. *Berthier.*

Geognostic and Geographic Situations.—Occurs in drusy cavities in micaceous clay-slate, with quartz and tile-ore; sometimes also with copper-pyrites, at Libethen, in Hungary.

Genus IV.—Blue Malachite, or Blue Copper.

Lazur Malachit, *Mohs.*

Prismatic. Streak and colour blue. Hardness = 3.5,—4.0. Sp. gr. = 3.5,—3.7,—3.9.

1. *Prismatic Blue Malachite.*

Prismatischer Lazur Malachit, *Mohs.* — Kupferlazur, *Werner.*—Cuivre carbonaté bleu, *Haüy.*

Specific Character.—Prismatic. Hemiprismatic $\frac{P}{2}$ = 116° 7′. $(\breve{P} + \infty)^2$ =59° 14′. Cleavage $(\breve{P} + \infty)^2$. Less distinct P—∞. Traces according to $\breve{P}r$ = 99° 32′.

Description.—Colours blue. Occurs regularly crystallized in concretions which are scopiform and stellular, radiated, and also curved lamellar; massive, globular, botryoidal, reniform, stalactitic, and cellular. Lustre ranges from shining to glimmering, and is vitreo-resinous. Fracture conchoidal. Ranges from transparent to translucent on the edges.

A variety, in dull and fine dusty particles, is named *earthy blue malachite,* while the other varieties are denominated *radiated blue malachite.*

Constituent Parts.—Copper 56.0, oxygen 12.5, carbonic acid 2.5. *Vauquelin.*

Geognostic and Geographic Situations.—Occurs in mineral veins and in beds in gneiss, mica-slate, greywacke, limestone, and red sandstone. It occurs at the Lead Hills, and in various English copper and lead mines.

* *Velvet Blue Copper.*

Kupfersammterz, *Werner.*

Description.—Occurs in bright blue, short capillary crystals. Lustre glistening and silky.

Geognostic and Geographic Situations.—It is a rare mineral; its only known locality is the Bannat, where it is associated with green malachite and brown iron-ore.

Genus V.—Emerald Malachite, or Dioptase.

Smaragd-Malachit, *Mohs.*

Rhombohedral. Colour lively emerald-green. Hardness = 5.0. Sp. gr. = 3.2,—3.4.

1. *Rhombohedral Emerald Malachite,* or *Dioptase.*

Rhomboedrischer Smaragd-Malachit, *Mohs.*—Kuferschmaragd, *Werner.*—Cuivre Dioptase, *Haüy.*

Specific Character.—Rhombohedral. Rhombohedron = 126° 17′. Cleavage rhombohedral. Streak green.

Description.—Colour emerald-green. Occurs regularly crystallized. Internally shining and pearly. Fracture conchoidal. Translucent, passing into semi-transparent.

Constituent Parts. — Oxide of copper 45.45, silica 43.18, water 11.36. *Vauquelin.*

Geognostic and Geographic Situations.—Is a very rare mineral, and has hitherto been found only in the land of the Kirgies in Tartary.

Genus VI.—Green Malachite.

Habronem-Malachit, *Mohs.*

Prismatic. Colour or streak bright green. Hardness = 2.5,—5. Sp. gr. = 3.5,—4.3.

1. *Prismatic Green Malachite,* or *Hydrous Phosphate of Copper.*

Prismatischer Habronem-Malachit, *Mohs.*—Phosphor Kupfererz, *Werner.*—Cuivre phosphaté, *Haüy.*

Specific Character.—Prismatic. Pyramid unknown. Cleavage, $\breve{P}r + \infty$. Imperfect. (Fig. 30.) Streak emerald-green. Hardness = 5.0. Sp. gr. = 4.0,—4.3.

Description.—Colours emerald-green, externally with a

blackish tarnish, or spotted black. Occurs regularly crystallized; in scopiform fibrous concretions; massive, reniform, botryoidal. Lustre ranges from shining to glimmering, and is resino-pearly. Fracture uneven. Opaque.

Constituent Parts.—Oxide of copper 62.85, phosphoric acid 21.68, water 15.45.

Geognostic and Geographic Situations.—Occurs at Virneberg, near Rheinbreitenbach, where it is disposed in veins in greywacke.

2. *Hemiprismatic Green Malachite*, or *Common Malachite.*

Hemiprismatischer Habronem-Malachit, *Mohs.*—Malachit, *Werner.*—Cuivre carbonaté vert, *Haüy.*

Specific Character.—Prismatic. Pyramid unknown. Cleavage, $P\bar{r} . P + \infty = 103°$ (nearly.) (Fig. 33, 30.) Streak grass or apple green. Hardness = 3.5,—4.0. Sp. gr. = 3.6,—3.7.

Description.—Colour green. Occurs regularly crystallized; in distinct concretions, which are scopiform, fibrous, angulo-granular, and wedge-shaped; also massive, disseminated, tuberose, stalactitic, reniform, botryoidal, fruticose, and cellular. Lustre ranges from shining to glimmering, and is silky. Fracture uneven, conchoidal, and even. Ranges from translucent to opaque.

Constituent Parts.—Copper 56.00, oxygen 14.00, carbonic acid 21.25, water 8.75. *Vauquelin.*

Geognostic and Geographic Situations.—Occurs in veins that traverse primitive, transition, and secondary rocks; also in beds, and disseminated through rocks of different kinds. The copper-mine of Sand-lodge, in Shetland, formerly afforded fine specimens of the fibrous varieties; and fine masses are sometimes found in the copper-mines of Cornwall.

3. *Prismatoidal Green Malachite or Muriate of Copper.*

Sallkupferez, *Werner.*—Cuivre muriaté, *Haüy.*—Atacamite.

Specific Character.—Prismatic. $P\breve{r} = 107° 10'$. $P + \infty = 67° 15'$. Cleavage $P\breve{r} + \infty$, perfect. Streak apple-green. Hardness = 3.0,—3.5. Sp. gr. = 4.0,—4.3.

Description.—Colour green. Occurs regularly crystallized; in radiated and granular concretions; massive, disseminated, and in scaly particles. Lustre shining, glistening, and pearly. Translucent on the edges.

Constituent Parts.—Oxide of copper 76.6, muriatic acid 10.6, water 12.8.—*Proust.*

Geognostic and Geographic Situations.—Occurs in veins along with ores of copper in Chili, in the form of grains and scales in alluvial sand in the Desert of Atacama in Peru. Is also found in fissures of some Vesuvian lavas.

Order V.—MICA.

Cleavage monotomous, very distinct. Hardness = 1.0, —4.5. Sp. gr. = 1.8,—3.4.

If metallic lustre, the sp. gr. is under 2.2. If no metallic lustre, the sp. gr. is above 2.2. If the streak is yellow, the sp. gr. is under 3.2. If the hardness is above 3.0, it is rhombohedral. If the sp. gr. is under 2.5, it is metallic.

Genus I.—Copper Mica.

Streak green. Hardness = 2.0. Sp. gr. = 2.5,—2.6.

1. *Rhombohedral Copper Mica*, or *Micaceous Arseniate of Copper.*

Rhomboedrischer Euchlor-glimmer, *Mohs.*—Kupferglimmer, *Werner.*—Cuivre arseniaté, *Haüy.*

Specific Character.—Rhombohedral. Rhombohedron = 68° 45′. Cleavage R—∞. Streak emerald,...apple-green.

Description.—Colour green. Occurs regularly crystallized; in granular concretions; massive, and disseminated. Internally splendent and pearly. Fracture uneven. Translucent and transparent. Sectile.

Constituent Parts.—Oxide of copper 39.0, arsenic acid 43.0, water 17.0. *Vauquelin.*

Geognostic and Geographic Situations.—Occurs in the copper-mines of Cornwall.

Genus II.—Uran-Mica, or Uranite.

Streak green,...yellow. Hardness = 2.0,—2.5. Sp. gr. = 3.0,—3.2.

1. *Pyramidal Uran Mica.*

Pyramidaler Euchlor-glimmer, *Mohs.*—Uranglimmer, *Werner.*—Uran Oxydé, *Haüy.*

Specific Character.—Pyramidal. Pyramid = 95° 13′; 144° 56′. Cleavage, P—∞.

Description.—Colours green and yellow. Occurs regularly crystallized; seldom massive, in scales, and in angulo-granular concretions. Lustre ranges from splendent to glistening, and is pearly. Transparent and translucent. Sectile. Not flexible. Easily frangible.

Constituent Parts.—Oxide of uran 59.37, phosphoric acid 14.63, water 14.90. *Berzelius.*

Geognostic and Geographic Situations.—Occurs in the copper and tin mines of Cornwall.

* *Uran Ochre.*—This is the yellow or reddish coloured, soft, earthy-looking, opaque mineral, occasionally associated with uran-mica, and also with uran-ore.

Genus III.—Cobalt Mica, or Red Cobalt.

Kobalt-Glimmer, *Mohs.*

Prismatic. Colour and streak red. Folia flexible. Hardness = 2.5. Sp. gr. = 2.9,—3.1.

1. *Prismatic Cobalt Mica,* or *Red Cobalt.*

Prismatischer Kobalt-Glimmer, *Mohs.*—Rother Erd Kobalt, *Werner.*—Cobalt arseniaté, *Haüy.*

Specific Character.—Prismatic. Pyramid unknown. Combination hemi-prismatic, $\frac{\bar{P}}{2}$. Cleavage, $\bar{P}r + \infty$. (Fig. 29.) Streak red.

Description.—Colours red, rarely grey, green, and brown. Occurs regularly crystallized; in stellular and scopiform fibrous concretions; massive, disseminated, in crusts, reniform, and botryoidal. Lustre ranges from shining to dull, and is pearly and resinous. Fracture earthy and conchoidal. Sectile. Ranges from translucent to opaque.

Constituent Parts.—Oxide of cobalt 39.0, arsenic acid 37.0, water 22.0. *Bucholz.*

Geognostic and Geographic Situations.—Occurs in veins in primitive and secondary rocks, and is met with in the coal-field around Edinburgh, and in the old lead-mines of Tyndrum in Perthshire.

* *Cobalt Ochre.*—There are two kinds of this mineral.

1. *Black Cobalt Ochre.*—Schwarz Erd Kobolt, *Werner.* Colours black and occasionally brown. Occurs botryoidal, reniform, and in crusts. Lustre glimmering or dull. Fracture earthy and conchoidal. Opaque. Streak shining and resinous. Very soft, and sometimes friable.

It is a compound of black oxide of cobalt, with arsenic and oxide of iron.

It occurs at Alderley Edge, Cheshire, in red sandstone.

2. *Brown and Yellow Cobalt Ochre.*—These differ from the preceding principally in colour, the tints being yellow and brown.

Genus IV.—Antimony Mica, or White Antimony.

Antimon-Glimmer, *Mohs.*

Hardness $= 2.5,—3.0$. Sp. gr. $= 5.5,—5.6$.

1. *Prismatic White Antimony.*

Prismatiches Antimon-glimmer, *Mohs.*—Weiss-spiessglaserz, *Werner.*—Antimoine Oxyde, *Haüy.*

Specific Character.—Prismatic. $Pr - 1 = 70° \ 32'$. $(\bar{P}_{,} +_{,}' \infty) = 136° \ 58$. Cleavage $(\bar{P}r + \infty)^2$, very perfect. $\breve{P}r + \infty$. Streak white or grey.

Description.—Colours white and grey. Occurs regularly crystallized; in concretions which are scopiform and stellular radiated, and also granular. Lustre shining and pearly adamantine. Translucent.

Constituent Parts.—Oxide of antimony 86.0, oxide of antimony and iron 3.0, silica 8.0. *Vauquelin.*

Geognostic and Geographic Situations.—It occurs in veins in primitive rocks, along with galena and grey and red antimony, in Bohemia, France, and Hungary.

* *Antimony Ochre.*—Spiesglanzocker, *Werner.*

Colours yellow, brown, and green. Occurs massive, disseminated, and in crusts. Dull. Earthy. Opaque. Very soft.

It always occurs in veins along with grey, and occasionally with red antimony, as in Cornwall.

Genus V.—Blue Iron, or Iron Mica.

Eisen-Glimmer, *Mohs.*

Prismatic. Streak white, grey,...blue. Hardness $=2.0$. Sp. gr. $=2.6,—2.7$.

1. *Prismatic Blue Iron, or Iron Mica.*

Prismatisches Eisen Glimmer, *Mohs.*—Vivianit, *Werner.* Fer Phosphate, *Haüy.*

Specific Character.—Prismatic. Pyramid unknown. Combination hemiprismatic, $\frac{\bar{P}}{2} = 119° 4'$. Cleavage $\bar{Pr} + \infty$. (Fig. 29.)

Description.—Colours blue and green. Occurs regularly crystallized; in scopiform and promiscuous fibrous concretions; massive, disseminated, and thinly coating. Lustre ranges from splendent to dull, and is pearly, inclining to adamantine. Ranges from transparent to opaque.

It sometimes occurs in a friable, or loosely cohering state, and is then composed of dull dusty particles, forming *earthy blue iron.*

Constituent Parts.—Protoxide of iron 47.5, phosphoric acid 32.0, water 20.0. *Klaproth.*

Geognostic and Geographic Situations.—The crystallized varieties are found in Cornwall, those in fibrous concretions in Greenland, and the earthy and friable in our peat mosses.

Genus VI.—Graphite.

Graphite Glimmer, *Mohs.*

Rhombohedral. Hardness = 1.0,—2.0. Sp. gr. = 1.8, —2.1.

1. *Rhombohedral Graphite.*

Rhomboedrischer Graphit-glimmer, *Mohs.*—Graphit, *Werner.*—Graphite, *Haüy.*

Specific Character.—Rhombohedral. Rhombohedron unknown. Combination dirhombohedral. Cleavage R — ∞. Metallic aspect. Streak black.

Description.—Colour dark steel-grey, inclining to iron-black. Occurs regularly crystallized; in granular concretions; massive and disseminated. Lustre ranges from splendent to glimmering, and is metallic. Fracture scaly foliated, uneven, conchoidal, and slaty. Opaque. Sectile.

Constituent Parts.—Carbon 92.0, iron 8.0. *Vauquelin.*

Geognostic and Geographic Situations.—It occurs in beds and imbedded masses, in primitive, transition, and secondary rocks. In Scotland it occurs in gneiss, in Glen Strath-Farrar, in Inverness-shire; in the coal-formation in Ayrshire; and in England, in transition rocks near Borrodale in Cumberland.

Genus VII.—Talc-Mica.

Talc-Glimmer, *Mohs.*

Rhombohedral. Prismatic. Streak white, grey,...green. Hardness = 1.0,—2.5. Sp. gr. = 2.7,—3.0. If the streak is green, the sp. gr. = 2.8 and less.

1. *Prismatic Talc-Mica, or Talc.*

Prismatischer Talc-glimmer, *Mohs.*

Specific Character.—Prismatic. P + ∞ = 120°. P + ∞

= 120°. Cleavage P — ∞. (Fig. 27.) Folia flexible. Streak white or green. Hardness = 1.0,—1.5. Sp. gr. = 2.7,—2.8.

Description.—Colours green, sometimes passing into greenish-black, also greenish-white, grey, and rarely blue. Occurs regularly crystallized; in granular, fibrous, and prismatic concretions; massive, disseminated, and in amygdaloidal pieces. Lustre ranges from splendent to dull, and is pearly or resinous. Fracture slaty, scaly foliated, earthy, and uneven. Ranges from translucent to opaque.

The dark green varieties, which are generally opaque, are named *chlorite;* of those the regularly crystallized are named *foliated chlorite;* the slaty, *chlorite slate;* the massive, scaly foliated, *common chlorite;* the massive in scaly foliated feebly adhering particles, *earthy chlorite;* and the dull earthy varieties met with in vesicular cavities in amygdaloid, *compact chlorite* or *green earth.* The white and paler green varieties are named *talc;* of these the most translucent, and possessing the highest degree of lustre, are denominated *common talc;* while the grey and green varieties, with slaty fracture and inferior lustre, and inferior translucency, are named *talc-slate* or *indurated talc.*

Constituent Parts.—1. *Foliated Talc.* Silica 62.0, magnesia 27.0, oxide of iron 3.5, alumina 1.5, water 6.0. *Vauquelin.*

2. *Chlorite Slate.* Silica 29.5, magnesia 21.4, oxide of iron, 23.4, alumina 15.6, water 7.4, lime 1.5. *Gruner.*

3. *Green Earth.* Silica 52.0, magnesia 6.0, oxide of iron 23.0, alumina 7.0, water 4.0, potash 7.5. *Vauquelin.*

Geognostic and Geographic Situations.—Occurs principally in primitive mountains, sometimes forming whole beds, as is the case with the varieties named chlorite-slate and talc-slate; other kinds, as common and earthy chlorite, occur disseminated, in veins, or dispersed through other minerals, as rock-crystal, felspar, &c., while the compact chlorite, or green earth, appears principally in secondary amygdaloid. The most beautiful variety of the species, the common talc, occurs chiefly in primitive rocks of limestone, mica-slate, &c. Examples of all the varieties are met with in the mountainous districts of Scotland and England.

* 1. *Native Magnesia, or Hydrate of Magnesia.*—Colours snow-white, greenish-white, and also grey. Occurs in prismatic concretions that point to the regular six-sided prism, also in granular concretions, and massive. Cleavage axotomous. Lustre shining and pearly. Semi-transparent in the mass, and transparent in single folia. Streak white, and affords on paper a polished pearly trace. Hardness = 1.0—1.5. Slightly elastic. Sp. gr. 2.13, *Bruce*; 2.336, *Brewster*. It is a pure hydrate of magnesia, the proportion of the constituent parts being, magnesia 70., water 30. Occurs in veins in serpentine, in the Shetland Islands, and in the serpentine of Portsoy, in Banffshire.

* 2. *Ophite*, or Precious Serpentine.—Colours pure leek-green, seldomer grass and oil green. Occurs massive and disseminated. Glistening and resinous lustre. Fracture flat conchoidal, inclining to splintery. Translucent, or translucent on the edges. Hardness 3.5. Sp. gr. = 2.5, —3.0.

* 3. *Nephrite.*—Colours green, grey, and white. Massive, and in rolled pieces. Dull or glimmering. Fracture splintery. Strongly translucent. Difficultly frangible. Hardness = 7.0. Sp. gr. = 2.9,—3.1.

It occurs imbedded in various primitive rocks in Persia, Egypt, and Germany.

4. *Steatite or Soapstone.*—Colours white, red, and yellow, with frequent dendritic markings of black. Occurs massive, in crusts, and in false crystals. Fracture splintery and uneven. Dull or glimmering. Translucent on the edges. Sectile. Soft. Feels very greasy. It occurs principally in serpentine, as in that of the Shetland Islands and Cornwall.

N. B. The *Pimelite* of authors appears to be steatite coloured with nickel or chrome.

* 5. *Figurestone or Algalmatolite.*—Colours grey, green, white, red, and brown. Massive. Dull or glimmering. Fracture conchoidal, or splintery and slaty combined. Translucent. Resinous in streak. Feels rather greasy. The finest varieties are brought to Europe from China.

* 6. *Magnesite.*—Colours white, grey, and cream-yellow. Occurs massive, tuberose, reniform, vesicular. Dull. Fracture conchoidal. Opaque. Hardness = 3.3. Sp. gr. = 2.881. Occurs in serpentine in Moravia, and in the Shetland Islands.

* 7. *Meerschaum.* —Colour white. Massive. Dull. Fracture earthy and conchoidal. Opaque or translucent on the edges. Very soft. Sectile. Adheres strongly to the tongue. Sp. gr. = 0.988,—1.279. Occurs in serpentine in Cornwall, Shetland Islands, &c.

* 8. *Lithomarge.*—Colours white, grey, blue, green, red, and yellow, and these are sometimes disposed in a veined, spotted, clouded, or striped manner. Occurs massive, disseminated, globular, and amygdaloid. Dull. Fracture earthy and flat conchoidal. Opaque. Shining in the streak. Soft. Sectile. Adheres strongly to the tongue. Feels fine and greasy. Sp. gr. —2.4. It occurs in trap, porphyry, and serpentine rocks, in Scotland and other countries.

* 9. *Mountain Soap.*—Colour blackish-brown. Massive. Dull. Fracture earthy. Opaque. Shining in the streak. Writes. Soft. Sectile. Adheres strongly to the tongue. Feels greasy. Occurs in secondary trap-rock, in the island of Skye.

* 10. *Bole.*—Colours brown, yellow, red, and black. Sometimes spotted, and dendritic. Massive and disseminated. Lustre glimmering. Fracture conchoidal. Feebly translucent. Soft. Feels greasy. Shining streak. Adheres to the tongue. Sp. gr. = 1.922. Occurs imbedded in secondary trap-rocks in Scotland.

2. *Rhombohedral Talc-Mica. Mono-axial Mica.*

Rhomboedrischer Talk-glimmer, *Mohs.*

Specific Character.—Regular six-sided prism, distinctly axotomous. Folia elastic. Streak white. Hardness = 2.0.—2.5. Sp. gr. = 2.8,—3.0.

Description.—Colours generally dark green and brown; sometimes also black. Lustre pearly, inclining to metallic on the terminal faces of the prism. From transparent to opaque. Streak white or grey. Sectile. Folia very flexible and elastic.

Constituent Parts.—Silica 42.50, alumina 16.05, magnesia 25.97, potash 7.55, oxide of iron 4.93. *Rose.*

Geognostic and Geographic Situations.—It occurs in primitive districts, and in beautiful brown-coloured crystals in the ejected masses found on the Monte Somma. The Siberian and the Vesuvian varieties have been most noticed by mineralogists.

Observation.—1. It is to be considered a rare mineral, when contrasted with the universally distributed Hemiprismatic or Common Mica.

2. It may be named *Mono-axial Mica*, from its exhibiting one axis of double refraction.

3. *Hemiprismatic Talc-Mica, or Common Mica. Di-axial Mica.*

Hemiprismatischer Talk-glimmer, *Mohs.*

Specific Character.—Oblique rhombic prism, with transverse sections of about 120° and 60°. Cleavage parallel to the base. Folia elastic. Streak white or grey. Hardness =2.0,—2.5. Sp. gr. =2.8,—3.0.

Description.—Colours white, grey, green, yellow, red, brown, and black. Lustre pearly. Transparent and translucent. Streak white or grey. Sectile. Flexible and elastic.

Constituent Parts.—1. From *Zinwald.* Silica 47.0. alumina 20.0, potash 14.50, oxide of iron 15.50, oxide of manganese 1.75, *Klaproth.* 2. *Red or Lepidolite Mica.* Silica 50.35, alumina 28.30, potash 9.04, oxide of manganese 1.23, fluoric acid and water 5.20, lithion 5.49, *Turner.* 3. *Zinnwald.* Silica 46.23, alumina 14.14, potash 4.90, oxide of iron 17.97, oxide of manganese 4.57, fluoric acid and water 3.73, lithion 4.21, *Gmelin.*

Geognostic and Geographic Situations.—Forms an essential constituent part of granite, gneiss, and mica-slate; abounds in many transition districts, and is frequent in sandstones of the secondary class. Hence its geographic distribution is universal.

Observation.—It exhibits two axes of double refraction, hence it may be named *Di-axial Mica.*

Genus VIII.—Pearl-Mica.

Perl-Glimmer, *Mohs.*

Rhombohedral. Hardness =3.5,—4.5. Sp. gr. = 3.0—3.1.

1. *Rhombohedral Pearl-Mica.*

Rhomboedrischer Perl-Glimmer, *Mohs.*

Specific Character.—Rhombohedral. Rhombohedron unknown. Combination dirhombohedral. Cleavage R—∞. Streak white or grey.

Description.—Colours white and grey. Occurs massive, in tables, and in granular concretions. Lustre splendent and pearly. Fracture foliated. Transparent to translucent. Streak white.

Constituent Parts.—Silica 37.0, alumina 40.5, lime 8.9, oxide of iron 4.50, natron 1.2, water 1.0. *Du Menil.*

Geognostic and Geographic Situations.—Occurs in beds, partly intermixed with mica and chlorite, at Sterzing, in the Tyrol.

Order VI.—SPAR.

Not metallic. Streak white, grey, brown, or blue. Hardness = 3.5,—7.0. Sp. gr. = 2.0,—3.7. If tessular, the sp. gr. = 3.0, and less. If rhombohedral, the sp. gr. = 2.2, or less, or the hardness = 6.0. If hardness = 4.0 and less, it is rhombohedral; or cleavage monotomous, and very distinct. If the hardness = 6.0, the lustre is pearly, the cleavage perfect, and sp. gr. less than 2.5, or more than 2.8. If sp. gr. above 3.3, the combination is hemi or tetarto prismatic; or the hardness = 6.0, but no adamantine lustre. If the sp. gr. = 2.4, and less, it is not without traces of form and cleavage.

Genus I.—Schiller-Spar.

Schiller-spath, *Mohs.*

Prismatic. Cleavage monotomous, perfect. Metallic pearly lustre. Hardness = 3.5,—6.0. Sp. gr. = 2.6,—3.4.

1. *Common Schiller-Spar.* *Diatomous*[1] *Schiller-Spar.*

Diatomer Schiller-spath, *Mohs.*—Schiller-stein, *Werner.* —Diallage Metalloide, *Haüy.*

Specific Character.—Prismatic. Pyramid unknown. Cleavage prismatoidal. Hardness = 3.5,—4.0. Sp. gr. = 2.6,—2.8.

Description.—Colours green, grey, and brown. Occurs in granular concretions; disseminated, and seldom massive. Lustre shining or splendent, and metallic pearly. Translucent on the edges, or opaque. Dull streak.

Constituent Parts.—Silica 43.90, magnesia 25.85, lime 2.64, oxide of iron 13.21, water 12.43. *Kohler.*

Geognostic and Geographic Situations.—Occurs imbedded in serpentine in the Shetland Islands, and in secondary trap-rocks in the middle district of Scotland.

2. *Bronzite.* *Hemiprismatic Schiller-Spar.*

Hemiprismatischer Schiller-spath, *Mohs.* — Blättriger Anthophyllite, *Werner.*—Diallage Metalloide, *Haüy.*

Specific Character.—Hemiprismatic. Pyramid unknown. Combination hemiprismatic. Cleavage $\frac{P}{2} . \breve{Pr} + \infty$. Less perfect, $\frac{\bar{Pr}}{2} . \bar{Pr} + \infty$. (Fig. 29, 35, 28.) Lustre metallic-pearly. Hardness = 4.0,—5.0. Sp. gr. = 3.0,—3.3.

Description.—Colours brown and grey. Occurs in granular concretions. Lustre shining. Cleavage sometimes appears fibrous. Translucent on the edges. Streak white.

Constituent Parts.—Silica 60.0, magnesia 27.5, oxide of iron 10.5, water 0.5. *Klaproth.*

Geognostic and Geographic Situations.—Occurs in serpentine in Shetland; in greenstone in the island of Skye, and near Portsoy; and in other parts of Scotland.

[1] From δια, through, and τεμνω, I cut; easily cleavable in one direction.

3. *Hypersthene.* *Prismatoidal Schiller-Spar.*

Prismatoidischer Schiller-spath, *Mohs.*—Paulit, *Werner.* —Hypersthene, *Haüy.*

Specific Character.—Prismatic. $P + \infty = 93° \; 0'$. Cleavage $\breve{P} + \infty$; $\breve{P}r + \infty$, perfect. $\bar{P}r + \infty$, only traces. (Fig. 29, 30, 28.) Lustre metallic-pearly. Hardness = 6.0. Sp. gr. = 3.3,—3.4.

Description.—Colours greenish and greyish black; also copper-red and brown. Occurs in granular and lamellar concretions and massive. Lustre shining. Opaque or feebly translucent.

Constituent Parts.—Silica 54.25, magnesia 14.00, alumina 2.25, lime 1.50, oxide of iron 24.50, water 1.0. *Klaproth.*

Geognostic and Geographic Situations.—Occurs in greenstone in the island of Skye; also in Banffshire and the Shetland Islands.

4. *Anthophyllite.* *Prismatic Schiller-Spar.*

Prismatischer Schiller-spath, *Mohs.*—Strahliger Anthophyllit, *Werner.*—Anthophyllite, *Haüy.*

Specific Character.—Prismatic. $P + \infty = 124° \; 30'$. Cleavage $\bar{P}r + \infty$. Rather less perfect, $\breve{P} + \infty$. $Pr + \infty$, least distinct. (Fig. 28, 30, 29.) Lustre metallic-pearly, inclining to vitreous. Hardness = 5.0,—5.5. Sp. gr. = 3.0,—3.3.

Description.—Colours yellowish grey and yellowish brown. Occurs crystallized in reed-like crystals; in scopiform and promiscuous radiated concretions; also massive. Translucent on the edges.

Constituent Parts.—Silica 56.0, alumina 3.0, magnesia 23.0, lime 2.0, oxide of iron 13.0, oxide of manganese 4.0. *Gmelin.*

Geognostic and Geographic Situations.—Occurs in primitive rocks near Drimnadrochit, in Inverness-shire.

Genus II.—Kyanite.

Disthen-spath, *Mohs.*

Prismatic. Cleavage prismatoidal, very distinct. Common pearly lustre. Hardness = 5.0,—7.0. Sp. gr. = 3.5,—3.7.

1. *Prismatic Kyanite.*

Disthen-spath, *Mohs.*—Kyanite and Rhatizit, *Werner*—Disthene, *Haüy.*

Specific Character.—Prismatic. Pyramid unknown. Combination tetarto-prismatic. Cleavage three faces, of varying degrees of perfection. Inclination of the most perfect to the less perfect = 100° 50′, to the least perfect = 106° 15′.

Description.—Colours blue and bluish green. Occurs regularly crystallized; in distinct concretions, which are granular and radiated, which latter are sometimes scopiform and stellular; also massive and disseminated. Lustre splendent and pearly. Translucent and transparent.

Constituent Parts.—Alumina 64.30, silica 34.33. *Arfwedson.*

Geognostic and Geographic Situations.—It occurs in primitive rocks in the Shetland Islands; also in Ba ffshire and Aberdeenshire.

Genus III.—Spodumene.

Hardness = 6.5,—7.0. Sp. gr. = 3.0,—3.1.

1. *Prismatic Spodumene.*

Spodumen, *Werner.*—Triphane, *Haüy.*

Specific Character.—Prismatic. $P + \infty = 93° 0$. Cleavage $P + \infty$. Somewhat more distinct, $\breve{P}r + \infty$. Not blue. (Fig. 30, 29.)

Description.—Colours greenish white and green. Occurs massive, and in large granular concretions. Lustre shining, glistening, and pearly. Fracture uneven. Translucent. Uncommonly easily frangible.

Constituent Parts.—Silica 66.49, alumina 25.30, lithion 8.85, oxide of iron 1.45. *Arfwedson.*

Geognostic and Geographic Situations.—Occurs in primitive rocks in Sweden and in Ireland.

Genus IV.—Prehnite.

Prehnite, *Werner* and *Haüy.*—Axotomer Triphan-spath, *Mohs.*

Hardness $= 6.0.$—$7.0.$ Sp. gr. $= 2.8,$—$3.0.$ Not blue.

1. *Axotomous Prehnite.*

Specific Character. — Prismatic. $P + \infty = 99° \; 30'$. Cleavage $P + \infty$. More distinct $P - \infty$. (Fig. 30, 27.)

Description.—Colours green, grey, and white. Occurs crystallized; in granular and in scopiform and stellular fibrous distinct concretions; massive and reniform. Lustre shining, glistening, and pearly. Fracture uneven. Ranges from transparent to translucent.

Constituent Parts.—Silica 43.60, alumina 23.00, lime 22.33, oxide of iron 2.00, water 6.40. *Thomson.*

Geognostic and Geographic Situations.—The more highly crystallized varieties have hitherto been found principally in primitive rocks, while the fibrous or less perfectly crystallized varieties occur principally in secondary trap-rocks. The secondary trap-rocks of Scotland afford many localities of this mineral.

Genus V.—Datolite.

Dystom-Spath, *Mohs.*

Prismatic. Cleavage obtained with difficulty. Internally, lustre resinous. Colour not blue. Hardness = 5.0,—5.5. Sp. gr. = 2.9,—3.0.

1. *Prismatic Datolite.*

Prismatischer Dystom-spath, *Mohs.*—Chaux Boratée Siliceuse, *Haüy.*

Specific Character.—Prismatic. Combination hemiprismatic. $\frac{\overset{\circ}{P}}{2} = 122° 0'$. $P + \infty = 77° 30$. Cleavage $P + \infty$, less perfect $\overset{\circ}{P}r + \infty$. (Fig. 30.)

Description.—Most frequent colours white and grey, seldom red, green, and yellow. Occurs regularly crystallized; also in granular, scopiform, and stellular, fibrous, and curved lamellar concretions; massive, reniform, and botryoidal. Lustre ranges from shining to dull, and is resinous or pearly. Fracture conchoidal, uneven, and earthy. Ranges from transparent to opaque.

The reniform and botryoidal varieties in fibrous concretions are named *Botryolite;* the earthy-looking and botryoidal *Earthy Botryoidal Datolite;* all the others *Common Datolite.*

Constituent Parts.—Silica 38.51, lime 35.59, boracic acid 21.34, water 4.60. *Du Menil.*

Geognostic and Geographic Situations.—Occurs in beds of magnetic iron-ore, subordinate to gneiss, near Arendal in Norway, but rarely in other parts of Europe.

Genus VI.—Zeolite.

Kouphon-Spath, *Mohs.*

Tessular, rhombohedral, pyramidal, prismatic. Hard

ness = 3.5,—6.0. Sp. gr. = 2.0,—2.5. If pyramidal, is distinctly axotomous. If hardness = 6.0, is tessular.

1. *Trapezoidal Zeolite, or Leucite.*

Trapezoidaler Kuphon-spath, *Mohs.*—Leuzit, *Werner.*—Amphigène, *Haüy.*

Specific Character.—Tessular. Cleavage hexahedral, dodecahedral, imperfect. Hardness = 5.5,—6.0. Sp. gr. = 2.4,—2.5.

Description.—Colours white and grey. Occurs regularly crystallized; in granular concretions, and in roundish grains. Lustre shining or glistening, and vitreo-resinous. Fracture flat conchoidal. Ranges from transparent to opaque.

Constituent Parts.—Silica 54.0, alumina 23.0, potash 22.0. *Klaproth.*

Geognostic and Geographic Situations.—Occurs in imbedded grains and crystals in older lavas, and associated with garnet, hornblende, quartz, glassy felspar, in the ejected masses of Monte Somma, and the extinct volcanoes on the Rhine, afford examples of leucite in both situations.

2. *Dodecahedral Zeolite,* or *Lapis Lazuli.*

Dodecaedrischer Kuphon-spath, *Mohs.*—Lazurstein, *Werner.* Lapis Lazuli, Sodalite, Hauyne, Spinellane, *Nosin,* of authors.

Specific Character.—Tessular. Cleavage dodecahedral, perfect. White, green, blue. Hardness = 5.5,—6.0. Sp. gr. = 2.25,—2.35.

Description.—Colours blue, green, white. Occurs in rhombohedral dodecahedrons, massive, and disseminated. Lustre shining and glistening, and vitreo-resinous. Fracture conchoidal or uneven. Streak white or grey. Translucent, or translucent on the edges.

Constituent Parts.—1. *Lapis Lazuli.* Silica 49.0, alumina 11.0, potash and soda 8.0, lime 16.0, oxide of iron 4.0, sulphuric acid 2.0, water a trace, *Gmelin.*—2. *Sodalite, from Greenland.* Silica 36.00, alumina 32.00, soda 25.00, muriatic acid 6.75, oxide of iron 0.25, *Eckeberg.*—3. *Hauyne.* Silica 33.48, alumina 18.87, potash 15.45, lime 12.00, oxide of iron 1.16, sulphuric acid 12.39, water 1.20, *Gmelin.*—4. *Spinellane.* Silica 43.0, alumina 29.5, soda 19.0, lime 1.5, oxide of iron 2.0, sulphuric acid 1.0, water 2.5, *Klaproth.*

Geognostic and Geographic Situations.—The Lapis Lazuli varieties of this species are found associated with primitive rocks in China, Persia, Siberia, and Bukhara. The Sodalite varieties occur in Greenland and Vesuvius; the Hauyne, in volcanic rocks in Vesuvius and on the banks of the Rhine; and the Nosin occurs at the Lake of Laach, near to Andernach.

3. *Hexahedral Zeolite,* or *Analcime.*

Hexaedrischer Kuphon-spath, *Mohs.*—Analcime, *Haüy.*—Kubizit, *Werner.*

Specific Character.—Tessular. Cleavage hexahedral, imperfect. Hardness = 5.5. Sp. gr. = 2.0,—2.2.

Description.—Colours white, grey, reddish white. Occurs regularly crystallized; in angulo-granular concretions, and massive. Lustre shining, glistening, and vitreo-pearly. Fracture uneven, or conchoidal. Ranges from transparent to translucent.

Constituent Parts.—Silica 55.07, alumina 22.22, soda 13.71, water 8.22. *Connell.*

Geognostic and Geographic Situations.—Occurs in secondary greenstone rocks in Fifeshire, Salisbury Craig, near Edinburgh, and in the same rock in other parts of Scotland.

4. *Paratomous Zeolite, or Cross-stone.*

Paratomer Kuphon-spath, *Mohs.*—Kreutzstein, *Werner.*—Harmotome, *Haüy.*

Specific Character.—Prismatic. Pyramid unknown. Cleavage P. $\breve{P}r+\infty$. $\bar{P}r+\infty$. (Fig. 29, 28.) Hardness =4.5. Sp. gr. =2.3,—2.4.

Description.—Colours white and grey, sometimes also yellow and red. Most frequently in regular crystals, which are frequently twin-crystals; seldom massive. Lustre shining, glistening, and vitreo-pearly. Fracture conchoidal and uneven. Translucent.

Constituent Parts.—Silica 47.04, alumina 15.24, baryta 20.85, lime 0.10, potash 0.88, water 14.92. *Connell.*

Geognostic and Geographic Situations.—Occurs in galena veins in the mines of Strontian in Argyleshire, in secondary trap-rocks in Dumbartonshire, and other parts of Scotland.[1]

5. *Rhombohedral Zeolite, or Chabasite.*

Rhomboedrischer Kuphon-Spath, *Mohs.* — Chabasie, *Haüy.*—Schabasit, *Werner.*

Specific Character.—Rhombohedral. Rhombohedron =94° 46′. Cleavage rhombohedral. Hardness =4.0, —4.5. Sp. gr. =2.0,—2.1.

[1] Another species is described by authors under the name *Phillipsite*, which differs from Crosstone in having sp. gr. = 2.0,—3.2, and containing in place of baryta, lime and potash. Hence this mineral is by some named *Lime Crosstone.*

Description.—Colour white. Occurs regularly crystallized, seldom massive. Externally splendent, internally glistening and vitreous. Fracture conchoidal and uneven. Translucent.

Constituent Parts.—Kilmalcolm. Silica 50.14, alumina 17.48, lime 8.47, potash with soda 2.58, water 20.83. *Connell.*

Geognostic and Geographic Situations.—Occurs in various secondary trap-rocks, especially amygdaloid, and is not unfrequent in the trap-rocks of Scotland.

6. *Diatomous Zeolite, or Laumonite.*

Diatomer Kuphon-spath, *Mohs.*—Laumonite, *Werner, Haüy.*

Specific Character.—Prismatic. $P+\infty=86^\circ 15'$. Combination hemiprismatic. Cleavage, $Pr+\infty$. More perfect, $\breve{Pr}+\infty$. (Fig. 28, 29.) Hardness unknown. Sp. gr. $=2.3,—2.4$.

Description.—Colour white. Occurs regularly crystallized, and in granular distinct concretions. Lustre shining, glistening, and pearly. When fresh is transparent; but on exposure to the atmosphere it very soon becomes opaque, and so soft as to yield to the pressure of the finger. Uncommonly easily frangible.

Constituent Parts.—Isle of Skye. Silica 52.04, alumina 21.14, lime 10.62, water 14.92. *Connell.*

Geognostic and Geographic Situations.—This mineral occurs in secondary trap-rocks in various parts of Scotland and Ireland.

7. *Prismatic Zeolite,* or *Mesotype.*

Prismatischer Kuphon-spath, *Mohs.*—Mesotype, *Haüy.*—Fibrous Zeolite, Natrolite.

Specific Character.—Prismatic. P = 143° 20′; 142° 40′; 53° 20′. P + ∞ = 91° 0′. Cleavage P + ∞, very perfect. (Fig. 30.) Hardness = 5.0—5.5. Sp. gr. = 2.0, —2.3.

Description.—Colours white, red, yellow, and yellowish-brown. Occurs regularly crystallized; in distinct concretions which are scopiform and stellular fibrous, granular, and curved lamellar; reniform, coralloidal, in plates and crusts. Lustre ranging from shining to dull, and is pearly. Fracture, in some varieties, coarse earthy. Ranges from nearly transparent to opaque.

The yellow and brown varieties, with striped colour delineations, and in which the fibrous and granular concretions are intersected by lamellar, are named *Natrolite;* those with earthy fracture and very soft, *Pearly zeolite;* while the other varieties are named *Fibrous zeolite.*

Constituent Parts.—Silica 54.46, alumina 19.70, soda 15.00, lime 1.61, water 9.83. Silica 47.21, alumina 25.60, soda 16.12, water 8.88, oxide of iron 1.35. Of these two analyses, the first is of common mesotype, the other of the natrolite-mesotype.

Geognostic and Geographic Situations.—Occurs principally in drusy cavities, or in veins in secondary trap-rocks in many places in Scotland. The natrolite is the rarest variety in Scotland.

Observation.—The natrolite variety is by some considered a distinct species.

8. *Prismatoidal Zeolite, or Stilbite.*

Prismatoidischer Kuphon-spath, *Mohs.*—Stilbite, *Haüy.*—Strahl-zeolith, *Werner.* Radiated zeolite.

Specific Character.—Prismatic. P = 190° 15′; 114° 0′;

96° 0′. $P + \infty = 94°$ 15′. Cleavage, $\breve{P}r + \infty$, very perfect; traces according to $\bar{P}r + \infty$. (Fig. 29.) Hardness = 3.5,—4.0. Sp. gr. = 2.0, —2.2.

Description.—Colour white, sometimes grey, yellow, or red. Occurs regularly crystallized; in granular, and in scopiform, and stellular prismatic concretions; also massive and globular. Externally splendent. Internally shining and pearly. Alternates from transparent to translucent.

Constituent Parts.—Silica 52.50, alumina 17.07, lime 11.52, water 18.45. *Thomson.*

Geognostic and Geographic Situations.—Occurs in secondary trap-rocks in many districts in Scotland.

9. *Hemi-prismatic Zeolite, or Heulandite.*

Hemi-prismatischer Kuphon-spath, *Mohs.*—Blätter Zeolith, *Werner.*—*Stilbite, Haüy.* Foliated Zeolite.

Specific Character.—Hemiprismatic. Pyramid unknown. Combination, hemiprismatic $\frac{\bar{P}}{2}$. Cleavage, $\breve{P}r + \infty$, very perfect. (Fig. 29.) Hardness = 3.5,—4.0. Sp. gr. = 2.0, —2.2.

Description.—Colours white, grey, red, and brown. Occurs regularly crystallized; in granular and lamellar distinct concretions. Fracture conchoidal. Externally, lustre splendent, shining, and vitreous; internally, shining and pearly. Ranges from transparent to translucent on the edges.

Constituent Parts.—Silica 59.95, alumina 16.87, potash 7.19, water 15.40. *Walmstadt.*

Geognostic and Geographic Situations.—Occurs in drusy

cavities in secondary trap-rocks in the Hebrides and Mainland of Scotland.

10. *Pyramidal Zeolite, or Apophyllite.*

Pyramidaler Kuphon-spath, *Mohs.*—Albin Ichthyophtralm, *Werner.*—Mesotype epointée, *Haüy.*

Specific Character.—Pyramidal. P = 104° 2′; 122° 0′. Cleavage P. — ∞, very perfect. [P + ∞] imperfect. Hardness = 4.5,—5.0. Sp. gr. = 2.2,—2.5.

Description.—Colour white. Occurs regularly crystallized; in straight and curved lamellar distinct concretions; massive and disseminated. Surface of the cleavage strongly iridescent. Externally splendent, the terminal planes of the prism pearly; internally glistening and vitreous. Ranges from transparent to translucent.

Constituent Parts.—From Iceland. Silica 50.76, lime 22.39, potash 4.18, fluoric acid trace, water 17.36. *Turner.*

Geognostic and Geographic Situations.—Occurs in secondary trap-rocks in the Hebrides and other parts of Scotland: the finest specimens are found in Iceland.

11. *Macrotypous Zeolite, Brewsterite.*

Specific Character.—Rhombohedral. R = 79° 29′ Cleavage rhombohedral but imperfect. Hardness = 4.0. Sp. gr. 20,—2.2.

Constituent Parts.—Silica 58.66, alumina 17.49, Strontia 8.32, baryta 6.75, lime 1.34, oxide of iron, 0.29, water 12.58. *Connell.*

Geognostic and Geographic Situations.—Found in the lead-mines of Strontian, in cavities of trap at Giant's Causeway, and in the lead mines of St Turpet near Freyburg in the Brisgau.

Genus VII.—Petalite.

Petalin-Spath, *Mohs.*

Prismatic. Hardness =6.0,—6.5. Sp. g. =2.4,—2.5.

1. *Prismatic Petalite.*

Prismatischer Petalin-spath, *Mohs.*

Specific Character.—Prismatic. Pyramid unknown. Cleavage P + ∞ = 95° (nearly). More perfect, according to $\bar{P}r + \infty$.

Description.—Colours red and white. Occurs massive; internally glistening, shining, and nearly pearly. Translucent. Brittle, and rather easily frangible.

Constituent Parts.—Silica 79.21, alumina 17.22, lithia 5.76. *Arfwedson.*

Geognostic and Geographic Situations.—It has hitherto been found only in a mine of black iron-ore in the island of Uton in Sudermania, associated with spodumene, felspar, quartz, mica, and tourmaline.

Observations.—It was in this mineral Arfwedson discovered the alkali named lithia, since met with in other species.

Genus VIII.—Felspar.

Rhombohedral pyramidal, prismatic. Hardness = 5.0, —6.0. Sp. gr. =2.5,—2.8. If hardness = 5.5, and less, the form is pyramidal, but cleavage not axotomous.

1. *Rhombohedral Felspar, or Nepheline.*

Rhomboedrischer feldspath, *Mohs.*—Nepheline, *Haüy and Werner.*

Specific Character.—Rhombohedral. R = 83° 55′. P = 139° 19′; 88° 6′. Combination di-rhombohedral. Cleavage, P — ∞. P + ∞, imperfect. Hardness = 6.0. Sp. gr. = 2.5,—2.6.

Description.—Colours white and grey. Occurs regularly crystallized, and massive. Externally splendent, internally shining and vitreous. Fracture conchoidal. Translucent, passing into transparent.

Constituent Parts.—Silica 43.36, alumina 33,49, soda 13.36, potash 7.13. *Gmelin.*

Geognostic and Geographic Situations.—Occurs in drusy cavities in granular foliated limestone, with meionite, Vesuvian, pleonaste, rhombohedral garnet, mica, &c. in the ejected masses on Monte Somma, near Naples.

2. *Prismatic Felspar, or Common Felspar. Potash Felspar.*

Prismatischer Feldspath, *Mohs.*—Orthoclase, *Breithaupt.* Ice-Spar.—Amazonstone.—Potash Felspar.

Hemiprismatic. $\frac{P}{2} = 126° 12'$. $(\breve{P} + \infty)^2 = 118° 52'$.

Cleavage $\frac{\breve{P}r}{2} = 64° 34'$, very perfect; $\bar{P}r + \infty$, nearly as perfect, but often interrupted; $(\breve{P} + \infty)^2$, imperfect. Hardness $= 6.0$. Sp. gr. $= 2.5$,—2.8.

Description.—Colours white, grey, green, blue, red, and brown; and sometimes with pearly opalescence, and beautiful changeability of colour. Occurs regularly crystallized; in distinct concretions, which are angulo-granular and lamellar; massive and disseminated. Lustre ranges from splendent to glistening, and even to dull, and is vitreo-pearly, or vitreous. Fracture conchoidal, splintery, slaty, and earthy. Ranges from transparent to opaque.

The transparent and translucent white-coloured varieties, with the silvery or pearly opalescence, are named *adularia,* also *moonstone;* the white and grey transparent varieties usually in small crystals, which are traversed by numerous rents, are named *glassy felspar*;[1] the translucent varieties, with various shades of colour, such as white and red, are the most abundant, and hence are named *common felspar*; the blue and green varieties are named *amazonstone*; the dark-grey varieties, with the beautiful changeability of colour, from Frederickswarn, are named *Norwegian Labrador felspar*; the feebly translucent compact varieties, with splintery fracture, are named *compact felspar*; the slaty varieties, with feeble lustre and translucency, are named *slaty felspar or clinkstone*; the varieties, in a comparatively loose state of aggregation, without lustre and transparency, according to the degrees of compactness, are named *porcelain earth*, *earthy felspar*, and *claystone*.

Constituent Parts.—Silica 64.0, alumina 20.0, lime 2.0, potash 14.0. *Vauquelin.*

Geognostic and Geographic Situations.—Potash-felspar is one of the most abundant minerals in nature, as it occurs in most of the principal rocks in primitive and transition mountains. It abounds in the alpine districts of Scotland, England, and Ireland, in granite, syenite, gneiss, porphyry, and quartz rocks.

3. *Tetarto-prismatic Felspar, or Albite. Soda Felspar.*

Cleavelandite. Tetartin, *Breithaupt.*—Soda felspar.—Some so called Adularias.

Specific Character.—Inclination of $\frac{\breve{\mathrm{Pr}}}{2}$ to $\mathrm{Pr} + \infty = 93°\ 20'$; of $r\,\frac{(P+\infty)^2}{2}$ to $l\,\frac{(P+\infty)^2}{2} = 121°\ 38'$. Combina-

[1] Glassy Felspar has been lately described as a distinct species under the name *Ryakolite.*

tion tetarto-prismatic. Cleavage $\frac{\breve{P}r}{2}$, $\bar{P}r+\infty$, very perfect. $\frac{(\breve{P}+\infty)^2}{2}$, generally less perfect. Hardness = 6.0. Sp. gr. = 2.6, — 2.68.

Description.—Colour generally white, sometimes grey, green, or red. Massive, and disseminated. Lustre upon faces of cleavage pearly, in other directions vitreous. Transparent, but more frequently translucent on the edges. Streak white. Brittle.

Constituent Parts.—From Chesterfield in America. Silica 70.68, alumina 19.80, soda 9.06, lime 0.23, *Stromeyer.* From Arendal. Silica 68.84, alumina 20.53, soda 9.12. Lime a trace. *Rose.*

Geognostic and Geographic Situations.—Forms a constituent part of the greenstone rocks so abundant around Edinburgh. Is associated with epidote and garnet in gneiss rocks at Arendal in Norway. Transparent crystals are found along with pearl-spar in the Tyrol. In the granite of the Mourne mountains in Ireland it is associated with common felspar.

4. *Polychromatic or Labrador Felspar. Lime Felspar.*

Polychromatischer feldspath, *Mohs.*

Specific Character.—Inclination of $-\frac{\breve{P}r}{2}$ to $\bar{P}r+\infty$ = 93° 20′. Combination tetarto-prismatic. Cleavage $-\frac{\breve{P}r}{2}$, very perfect; $\bar{P}+\infty$, less perfect. Hardness = 6.0. Sp. gr. 2.69,— 2.76.

Description.—Colour grey, with splendent changeability of colour, exhibiting patches of brilliant blue, green, yellow, red, and pear-grey colours. Occurs massive and disseminated. Lustre pearly on the perfect cleavage faces. The beautiful changeability of colour seen to greatest advantage upon particular faces of the mass. Translucent, or nearly opaque.

Constituent Parts.—Silica 55.75, alumina 26.50, lime 11.0, soda 4.0, oxide of iron 1.25, water 0.50.

Geognostic and Geographic Situations.—This beautiful mineral was first noticed on the coast of Labrador as a constituent part of syenite, hence its name Labradorite. Sir Charles Giesecké found it also in Greenland. It has also been met with in different parts of Europe.

5. *Pyramidal Felspar, or Scapolite. Meionite.*

Pyramidaler Feldspath, *Mohs.*—Meionite Scapolith, Schmelzstein. *Werner.*—Paranthine, Meionite, Wernerite, Dipyre, *Haüy.*

Specific Character.—Pyramidal. Pyramid $= 136° 7'$; $63° 48'$. Cleavage $P - \infty$. More perfect, $P + \infty$. $[P + \infty]$. Hardness $= 5.0, - 5.5$. Sp. gr. $= 2.5, - 2.8$.

Description.—Colours white, grey, green, red, and black. Occurs regularly crystallized; in distinct concretions, which are scopiform, fibrous, or radiated, and angulo-granular; massive and disseminated. Lustre ranges from splendent to glimmering, and is pearly, resino-vitreous, and resino-pearly. Fracture conchoidal, uneven. Ranges from transparent to opaque.

The white and more transparent and highly crystallized varieties are named *Meionite,* while the others have received the names *Scapolite, Paranthine, Wernerite, Dipyre, and Schmelzstein.*

Constituent Parts.—1. *Meionite.* Silica 45.53, alumina 32.73, lime 24.25, potash and soda 1.82, protoxide of iron

0.18. *Stomeyer*.—2. *Scapolite*. Silica 43.83, alumina 35.43, lime 18.96, water 0.03. *Nordenskiold*.—3. *Wernerite*. Silica 50.25, alumina 30.00, lime 10.45, potash and soda 2.00, protoxide of iron 4.45, water 2.85. *John*.—4. *Dipyre*. Silica 60.0, alumina 24.0, lime 10.0, water 2.0. *Vauquelin*.

Geognostic and Geographic Situations.—The *meionite* varieties are found in drusy cavities in granular foliated limestone, along with nepheline, augite, mica, pleonaste, garnet, and calcareous spar, on Monte Somma, near Naples; *scapolite* and *Wernerite* varieties occur in beds of magnetic ironstone and iron-pyrites, in gneiss, along with quartz, felspar, mica, hornblende, epidote, garnet, augite, &c. at Arendal in Norway; the *paranthine* varieties in limestone quarries in Wermeland; the *dipyre* varieties in the torrent of Mauleon in the Western Pyrenees.

Observations.—Breithaupt describes a felspar under the name *Perecline*, nearly allied to albite, but distinguished from it by its inferior specific gravity, it being only 2.54,—2.56; and Rose describes another species under the name *Anorthite*, but the limited nature of this article prevents us noticing them more particularly.

Genus IX.—Augite.

Prismatic. Lustre not metallic-pearly. Hardness = 4.5,—7.0. Sp. gr. = 2.7,—3.5. If the hardness is above 6.0, the sp. gr. = 3.2, and more. If the sp. gr. is under 3.2, the cleavage is very perfect, and, in obliquely intersecting directions, parallel to the axis.

1. *Augite*. *Paratomous Augite*.

Paratomer Augit, *Mohs*.—Augit, *Werner*.—Pyroxene, *Haüy*.

Specific Character.—Prismatic. Pyramid = 152° 12′;

120°; 61° 2′. $P + \infty = 51°\ 19'$. Combination hemiprismatic . $\frac{P}{2} = 120°$. Cleavage, $(\breve{P}r + \infty .)^3 = 87°\ 42'$, $\breve{P}r + \infty$. $\bar{P}r + \infty$. Sometimes $\frac{\bar{P}}{2}$. (Fig. 32, 29, 28, 38.)

Hardness = 5.0,—6.0. Sp. gr. = 3.2,—3.5.

Description.—Colours green, black, and brown; also grey and white. Occurs regularly crystallized; in granular and fibrous concretions; massive and disseminated. Lustre ranges from splendent to glimmering, and is vitreo-resinous, resino-pearly, and resinous. Fracture conchoidal and uneven. Ranges from transparent to translucent on the edges.

Those varieties in which the colours are white and pale green, generally crystallized, with a vitreous external, and pearly internal lustre, and translucent, are named *Diopside, Musite, Alalite,* and by some *Baikalite;* those, again, in which the colours are darker green and muddy grey, less frequently crystallized, but disposed in straight, lamellar, and granular concretions, with a shining, vitreous, or pearly lustre, and translucent on the edges, are named *Sahlite, Pyrogome, Fassaite, Malacolite,* and by some *Baikalite;* other varieties, in which the colours are black and dark green, with conchoidal and uneven fracture, resinous lustre, and opaque or faintly translucent on the edges, are named *Conchoidal Augite and Common Augite;* those varieties in which the colours are principally leek-green and greenish-black, generally in loosely aggregated angulo-granular concretions, with a shining vitreo-resinous lustre, uneven fracture, and ranging from translucent to opaque, are named *Coccolite and Granular Augite;* and, lastly, the fibrous varieties have been described as *Amianthus.*

Constituent Parts.—1. *White Varieties.* Silica 54.83, lime 24.76, magnesia 18.55, alumina 0.28, protoxide of iron 0.99. *Bonsdorf.*—2. *Green Varieties.* Silica., 54.08, lime 23.47, magnesia 11.49, protoxide of iron 10.02, manganese 0.51. *Rose.*—3. *Black Varieties.* Silica 53.36, lime 22.19, magnesia 4.19, protoxide of iron 17.38. oxide of manganese 0.09. *Rose.*

Geognostic and Geographic Situations.—The diopside varieties are found imbedded in serpentine, and associated with magnetic iron-ore in Piedmont: the sahlite in beds of primitive trap, limestone, and magnetic iron-ore, subordinate to gneiss and mica-slate in Scotland, Ireland, and Scandinavia; the conchoidal augite and common augite occur principally in secondary trap-rocks; and the coccolite and granular augite in the iron mines of Arendal, in Norway, in gneiss.

2. *Hornblende. Hemiprismatic Augite.*

Hemi-prismatischer Augit-spath, *Mohs.*—Hornblende, &c., *Werner.*—Amphibole, *Haüy.*

Specific Character.—Prismatic. Pyramid = 151° 8′; 148° 39′; 42° 22′ . $P + \infty = 87°\ 11'$. Combination hemiprismatic, $\frac{P}{2} = 148°\ 39'$. Cleavage, $(\breve{Pr} + \infty)^3 =$ 124° 34′. Less distinct, $\breve{Pr} + \infty$. $\bar{Pr} + \infty$. (Fig. 32, 29, 28.) Hardness = 5.0,—6.0. Sp. gr. = 2.7,—3.2.

Description.—Colours green, white, black, grey, blue, and brown. Occurs regularly crystallized; in fibrous, radiated, and granular distinct concretions. Lustre ranges from splendent to feebly glimmering, and lustre vitreo-pearly, pearly, or vitreous, and vitreo-resinous. Fracture conchoidal, uneven, and slaty. Alternates from transparent to opaque.

The varieties with dark green and black colours, in granular and fibrous concretions, with a lustre varying from pearl to pearly-vitreous, and whose transparency ranges from transparent to translucent on the edges, are named *Hornblende* and *Carinthine;* the varieties with light green, and also greenish-grey, and sometimes brown and yellow colours, rarely crystallized (and then generally in reed-like crystals), more frequently massive, and in radiated, fibrous, and granular concretions, with a pearly or vitreous, splendent, shining, or glistening lustre, and transparency varying from transparent to opaque, are named *Actynolite* and *Calamite;* the white and blue varieties, disposed in fibrous, radiated, and granular concretions, with a lustre which is shining, or glistening and pearly, or vitreo-pearly, and ranging from translucent to translucent on the edges, are named *Tremolite;* other varieties, in which the colours are white, green, yellow, blue, and brown, and disposed in flexible, shining, silky, fibrous concretions, are named *Flexible Asbestus* or *Amianthus;* others, which are of a white colour, or grey colour, and disposed in minute promiscuous fibrous concretions, and so light as to swim in masses in water, are named *Mountain Cork;* those in which the colours are generally green, and disposed in straight, shining, pearly, rigid, fibrous concretions, are named *Rigid or Common Asbestus;* and, lastly, those varieties in which the colour is wood-brown, and in general aspect much resembling fossil-wood, are named *Rockwood* or *Ligneous Asbestus.*

Constituent Parts.—White. Silica 60.31, magnesia 24.23, lime 13.66, alumira 0.26, protoxide of iron 0.15, fluoric acid 0.94, water and foreign matter 0.10. *Bonsdorf.*—*Green.* Silica 46.26, magnesia 19.03, lime 13.96, alumina 14.48, protoxide of iron 3.43, protoxide of manganese 0.36, fluoric acid 1.60, water, &c. 1.04. *Bonsdorf.*—*Black.* Silica 45.69, magnesia 18.79, lime 13.85, alumina 1.18, protoxide of iron 7.32, fluoric acid 1.50. *Bonsdorf.*

Geognostic and Geographic Situations.—The *hornblende* varieties abound in primitive and transition rocks, and are also met with in secondary rocks; the *actynolite* and *tremolite* varieties scarcely occur in secondary rocks, being confined principally to those of the primitive class; while the *asbestine* varieties are principally met with in the serpentine rocks of primitive and transition mountains. Numerous localities of this species occur in Scotland, England, and Ireland.

3. 2. *Epidote. Pistacite. Prismatoidal Augite.*

Prismatoidischer Augit-spath, *Mohs.*—Epidote, *Haüy.*

Specific Character.—Prismatic. Pyramid unknown. Combination hemiprismatic. Cleavage, two faces, of which one is more distinct than the other. Incidence = 114° 37′. Hardness = 6.0,—7.0. Sp. gr. = 3.2,—3.5.

Description.—Colours green and grey. Occurs regularly crystallized; in granular, fibrous, and prismatic concretions. Lustre ranges from splendent to glimmering, and is resino-pearly. Fracture conchoidal, uneven, splintery, and sometimes nearly earthy. Ranges from transparent to translucent on the edges. Brittle, and easily frangible.

The green varieties are named simply epidote, or *pistacite;* while the grey and less perfectly crystallized varieties are named *zoisite.*

Constituent Parts.—1. *Pistacite, or Green Varieties.*

silica 37.0, alumina 27.0, lime 14.0, oxide of iron 17.0, oxide of manganese 1.5. *Descostes.*—2. *Grey Variety, or Zoisite.* Silica 45.0, alumina 29.0 lime 21.0, oxide of iron 3.0. *Klaproth.*

Geognostic and Geographic Situations.—It occurs principally in primitive rocks, such as gneiss, mica-slate, syenite, &c. Inverness-shire, Ross-shire, the Shetland Islands, and other parts of Scotland, afford many varieties.

4. *Wollastonite. Prismatic Augite.*

Prismatischer Augit-spath, *Mohs.*—Schaalstein, *Werner.*—Spath en Tables, *Haüy.* Table Spar.

Specific Character.—Prismatic. Pyramid unknown. Cleavage, $P + \infty = 105°$ (nearly). $\breve{Pr} + \infty$. $\bar{Pr} + \infty$. Sp. gr. = 2.7,—2.9. Hardness = 4.5,—5.0.

Description.—Colour white. Occurs in granular and lamellar concretions; also massive and disseminated. Lustre ranges from shining to glistening, and is pearly-vitreous. Fracture splintery. Translucent. Brittle.

Constituent Parts.—Silica 51.45, lime 47.41, oxide of iron 0.40. *Stromeyer.*

Geognostic and Geographic Situations.—Occurs in primitive rocks in the Bannat, in Norway, Finland, and in rocks of the same description in the island of Ceylon.

Genus X.—Azure-Spar.

Colour blue. Prismatic. Hardness = 5.0,—6.0. Sp. gr. = 2.8,—3.1.

1. *Prismatic Azure-Spar.*

Lazulit, *Werner.*—Lazulite, *Haüy.*—Prismatischer Lazure-Spath, *Mohs.*

Specific Character.—Prismatic. Cleavage P + ∞ (imperfect). Colour lively. Streak white. Hardness = 5.0, —5.5. Sp gr. = 3.0,—3.1.

Description.—Colour blue. Occurs massive. Fracture uneven. Opaque. Very feebly translucent on the edges. Easily frangible.

Constituent Parts.—Phosphate of lime 41.81, alumina 35.73, magnesia 9.34, silica 2.10, protoxide of iron 2.64, water 6.06. *Fuchs.*

Geognostic and Geographic Situations.—Occurs imbedded in quartz in the district of Vorau, in Stiria.

2. *Prismatoidal Azure-Spar, or Blue Spar.*

Blauspath, *Werner.*—Feldspath bleu, *Haüy.*

Specific Character.—Prismatic. Cleavage prismatoidal, imperfect. Colour pale. Streak white. Hardness = 5.5, —6.0. Sp. gr. = 3.0,—3.1.

Description.—Colour blue. Occurs massive and disseminated. Lustre glistening. Fracture splintery. Feebly translucent. Yields a greyish-white streak.

Constituent Parts.—Alumina 34.50, phosphoric acid 43.32, magnesia 13.56, silica 6.50, oxide of iron 0.80, water 0.50. *Brandes.*

Geognostic and Geographic Situations.—Occurs in primitive rocks at Krieglach in Stiria.

3. *Uncleavable Azure-Spar, or Calaite—Mineral Turquois.*

Specific Character.—No cleavage. Hardness = 6.0. Sp. gr. = 2.8,—3.0.

Description.—Massive. Disseminated. Blue. Streak white. Feebly translucent on the edges, or opaque. Internally dull. Fracture conchoidal.

Constituent Parts.—According to Berzelius it consists of phosphate of alumina and phosphate of lime with silica, the oxides of iron and copper, and a little water. John, the Russian chemist, on the contrary, found it composed of alumina 73.0; copper 4.5; oxide of iron 4.0; water 18.0.

Geognostic and Geographic Situations.—This beautiful mineral occurs in small veins in a slate-clay in Persia, where it is much prized as an ornamental stone.

Observation.—The Bone-Turquois is not true Calaite, but bone coloured with oxide of copper.

Order VII.—GEM.

Not metallic. No metallo-adamantine lustre. Streak white or grey. Hardness = 5.5,—10.0. If hardness = 6.0, and less, it is tessular, and sp. gr. = 3.1, and more; or sp. gr. = 2.4, and less, and no traces of form or cleavage. If sp. gr. is less than 3.8, there is no pearly lustre.

Genus I.—Andalusite.

Prismatic. Cleavage perfect, not prismatoidal. Hardness = 7.5. Sp. gr. = 3.0,—3.2.

1. *Prismatic Andalusite.*

Prismatischer Andalusit, *Mohs.*—Andalusit, *Werner.*—Feldspath apyre, *Haüy.*

Specific Character.—Prismatic. Pyramid unknown. Cleavage, $P + \infty$. $\mathring{P}r + \infty$. $\bar{P}r + \infty$. (Fig. 30, 29, 28.)

Description.—Colours red and grey. Occurs regularly crystallized and massive. Lustre shining, glistening, and vitreous. Fracture uneven. Feebly translucent.

Constituent Parts.—Alumina 60.5, silica 36.5, oxide of

iron 4.0. *Bucholz.*—Alumina 52.0, silica 28.0, oxide of iron 2.0, potash 8.0. *Vauquelin.*

Geognostic and Geographic Situations.—Occurs in gneiss in the Shetland Islands, and in slate rocks in Banffshire.

* *Fibrolite.*—Bournon.

Prismatic. $P + \infty = 100°$. Cleavage imperfect. Hardness more considerable than quartz, and sp. gr. = 3.214.

Constituent Parts.—According to Chenevix consists of silica 33.0, alumina 46.0, oxide of iron 13.0. Occurs in the Carnatic and in China.

* *Chiastolite.* Holspath, *Werner.*—Macle, *Haüy.*

Prismatic. $P + \infty = 84° 48$. $\bar{Pr} = 120$ nearly. Cleavage, $P + \infty$. $\overset{\circ}{Pr} + \infty$. $\bar{Pr} + \infty$. All of them imperfect. Hardness = 5, — 5.5. Sp. gr. = 2.9, — 3.0.

Constituent Parts.—Silica 68.49, alumina 30.17, magnesia 4.12, oxide of iron 2.7, water 0.27.—*Landgrabe.*

Geognostic and Geographic Situations.—Occurs in clay-slate near Keswick in Cumberland, and near Balahulish in Argyllshire.

Genus II.—Corundum.

Tessular, rhombohedral, prismatic. Hardness = 8.0. — 9.0. Sp. gr. = 3.5, — 4.3. If prismatic, the sp. gr. = 3.65, and more, and hardness = 8.5. If colour red, or brown, and sp. gr. = 3.7, and more, the hardness = 9.0.

1. *Dodecahedral Corundum, or Spinel.*

Dodecaedrischer Corund, *Mohs.*—Spinel and Zeilanit, *Werner.*—Spinelle and Pleonaste, *Haüy.*

Specific Character.—Tessular. Cleavage, octahedral,

but obtained with difficulty. Hardness = 8.0. Sp. gr. = 3.5,—3.8.

Description.—Colours red, and sometimes green, black, blue, yellow, brown, and white. Occurs regularly crystallized. Lustre splendent and vitreous. Fracture conchoidal. Ranges from transparent to translucent on the edges.

The dark green and black varieties, which are only translucent on the edges, are named *ceylanite;* the other varieties *spinel* or *spinel-ruby.*

Constituent Parts.—*Red Spinel.* Alumina 74.50, silica 15.56, magnesia 8.25, oxide of iron 1.56, lime 0.75. *Klaproth.*

Geognostic and Geographic Situations.—The ceylanite varieties occur in the sand of rivers in Ceylon; the others also in Ceylon, Pegu, and other countries.

2. *Octahedral Corundum or Automolite.*

Octaedrischer Corund, *Mohs.*—Automolite, *Werner.*—
Gahnite. Spinelle Zincifere, *Haüy.*

Specific Character.—Tessular. Cleavage, octahedral, and easily obtained. Hardness = 8. Sp. gr. = 4.1,—4.3.

Description.—Colour muddy dark-blue, passing into blackish-green. Occurs regularly crystallized. External lustre pearly, inclining to semimetallic, internally shining and resinous. Fracture conchoidal. Opaque or faintly translucent on the edges.

Constituent Parts.—Alumina 60.00, oxide of zinc 24.25, oxide of iron 9.25, silica 4.75. *Eckeberg.*

Geognostic and Geographic Situations.—Occurs imbedded in talc, and associated with lead-glance at Fahlun in Sweden, also in New Jersey in the United States of America.

3. *Rhombohedral Corundum. Sapphire. Oriental Ruby, &c.*

Rhomboedrischer Corund, *Mohs.*

Specific Character.—Rhombohedral. Rhombohedron $= 86° 6'$. Cleavage $R - \infty$. More perfect R. Hardness $= 9.0$. Sp. gr. $= 3.0, —4.05$.

Description.—Colours blue, red, grey, white, green, yellow, brown, and black. Occurs regularly crystallized, also massive. Lustre ranges from splendent to glimmering, and vitreous, or vitreous sometimes inclining to adamantine. Fracture conchoidal and uneven. Ranges from transparent to feebly translucent on the edges.

The transparent blue varieties are named *sapphire ;* the transparent red varieties, *oriental ruby ;* the massive, nearly opaque, bluish-grey and greyish-black varieties, *emery ;* the translucent, massive, and crystallized varieties, *common corundum ;* and the brown faintly translucent varieties, *adamantine spar.*

Constituent Parts.—*Sapphire.* Alumina 98.5, oxide o iron 1.0, lime 0.5. *Klaproth.*—*Emery.* Alumina 86.0, silica 3.0, oxide of iron 4.0. *Tennant.*

Geognostic and Geographic Situations.—The finest sapphires and oriental rubies are found in alluvial soil in Ceylon and Pegu, and other eastern countries; the emery occurs in primitive talc-slate in Saxony; the common corundum and adamantine spar in granite, syenite, and other similar rocks in India and China.

4. *Prismatic Corundum, or Chrysoberyl.*

Prismatischer Corund, *Mohs.*—Krysoberyll, *Werner.*—Cymophane, *Haüy.*

Specific Character.—Prismatic. $\breve{\mathrm{P}}\mathrm{r} = 119° 46'$; $(\breve{\mathrm{P}} + \infty)^3$

$= 70° 41'$. Cleavage $= \breve{P}r + \infty$. Less perfect $Pr + \infty$. (Fig. 29, 28.) Hardness = 8.5. Sp. gr. = 3.65, — 3.8.

Description.—Colour asparagus green, and often exhibits a milk-white opalescence. Occurs regularly crystallized, and in blunt edged pieces. Lustre splendent and resino-vitreous. Fracture conchoidal. Semi-transparent and transparent.

Constituent Parts.—Alumina 76.75, glucina 17.79, protoxide of iron 4.49, moisture 0.48. *Thomson.*

Geognostic and Geographic Situations.—Occurs imbedded in granite veins in America, in alluvial soil in Ceylon, and apparently in the granite of Aberdeen.

Genus III.—Diamond.

Tessular. Hardness = 10. Sp. gr. = 3.4, — 3.6.

1. *Octahedral Diamond.*

Specific Character.—Tessular. Cleavage octahedral, very perfect.

Description.—Its colours are more numerous than in most other minerals, and of the various tints the grey and white are the most frequent; it exhibits also beautiful red, yellow, and blue varieties. Occurs regularly crystallized, and in roundish grains. The lustre splendent and adamantine. Fracture conchoidal. Transparent and semi-transparent.

Constituent Parts.—Consists of pure carbon, and is completely volatilized at a temperature of 14°, Wedgewood, affording with oxygen, carbonic acid gas.

Geognostic and Geographic Situations.—It has hitherto been found principally in alluvial soil, in the warmer regions of the earth, as Brazil, Borneo, the Peninsula of India, and lately is reported to have been discovered in Siberia.

Genus IV.—Topaz.

Prismatic. Cleavage axotomous. Hardness = 8. Sp. gr. = 3.4,—3.6.

1. *Prismatic Topaz.*

Prismatischer Topaz, *Mohs.*—Topaz. Physalit. Piknit, *Werner.*—Silice Fluatée Alumineuse, *Haüy.* Schorlite.

Specific Character.—Prismatic. Pyramid = 141° 7′; 101° 52′; 90° 55′. P + ∞ = 124° 19′. Combination, sometimes with different faces on opposite ends. Cleavage, P — ∞, and very perfect. (Fig. 27.)

Description.—Colours yellow, green, blue, red, grey, white. Occurs regularly crystallized; also in prismatic and granular concretions, and massive. Lustre ranges from splendent to glistening, and is vitreous and resinous. Fracture conchoidal and uneven. Ranges from transparent to translucent on the edges. Easily and uncommonly easily frangible.

The highly crystallized and transparent varieties are named *precious topaz;* those in prismatic distinct concretions, with a slight degree of translucency on the edges, and which are uncommonly easily frangible, *schorlite or schorlous topaz;* and those in coarse granular concretions, with a low degree of lustre, and feeble translucency on the edges, *physalite.*

Constituent Parts.—1. *Precious Topaz.* Alumina 57.45, silica 34.24, fluoric acid 7.75. *Berzelius.*—2. *Schorlous Topaz.* Alumina 51.00, silica 38.43, fluoric acid 8.84. *Berzelius.*—3. *Physalite.* Alumina 57.74, silica 34.36, fluoric acid 7.77. *Berzelius.*

Geognostic and Geographic Situations.—Precious topaz occurs in alluvial soil in the upper parts of Aberdeenshire, in primitive rocks in Cornwall, in granite in the Mourne Mountains in Ireland; schorlous topaz is said also to occur in Aberdeenshire; and the physalite in granite, at Finbo in Sweden, and at Altenberg in Saxony. The Uralian and Altain Mountains in Asiatic Russia, and those of Brazil in the New World, afford rich and beautiful varieties of this gem.

Genus V.—Emerald.

Rhombohedral. Prismatic. Cleavage rhombohedro-axotomous and peritomous, or very perfect prismatoidal. Hardness = 7.5,—8.0. Sp. gr. 2.6,—3.2.

1. *Prismatic Emerald or Euclase.*

Euclase, *Werner*.—Prismatischer Smaragd, *Mohs*.—Euclase, *Haüy*.

Specific Character.—Prismatic. Hemiprismatic. Inclination of $\frac{\bar{P}r}{2}$ to Pr + ∞ = 130° 8′. Cleavage, $\mathring{P}r + \infty$, very distinct. $\frac{\bar{P}r}{2}$, rather perfect. Hardness = 7.5. Sp. gr. = 2.9, — 3.2.

Description.—Colours green, white, and blue. Occurs regularly crystallized. Lustre splendent. Fracture conchoidal. Transparent and translucent. Very easily frangible.

Constituent Parts.—Silica 43.22, alumina 30.56, glucina 21.78, oxide of iron 2.22, oxide of tin 0.70. *Berzelius*.

Geognostic and Geographic Situations.—This beautiful and rare mineral has been hitherto found only in Peru and Brazil.

2. *Rhombohedral Emerald. Beryl.*

Rhomboedrischer Smaragd, *Mohs*.—Emeraude, *Haüy*.

Specific Character.—Rhombohedral, R = 104° 40′.

2 (R) = 138° 41′; 89° 45′. P = 151° 9′; 59° 47′ Combination dirhombohedral. Cleavage, R — ∞; less perfect, P + ∞. Hardness = 7.5, — 8.0. Sp. gr. = 2.6, —2.8.

Description.—Colours green, blue, yellow, and grey. Occurs regularly crystallized, and in thin prismatic concretions. Lateral planes longitudinally streaked or smooth. Lustre ranges from splendent to glistening, and is vitreous. Fracture conchoidal. Ranges from transparent to translucent on the edges. Easily frangible.

The varieties, with emerald-green colours, in short, smooth, transparent, and translucent prisms, with rough terminal planes, are named *precious emerald*, while the others, in which the colours are green, blue, yellow, and grey, and crystallized in long longitudinally streaked prisms, are named *beryl*, or *common emerald*.

Constituent Parts.—1. *Precious Emerald.* Silica 68.50, alumina 15.75, glucina 12.50, lime 0.25, oxide of chrome 0.30, oxide of iron 1.00. *Klaproth.*—2. *Beryl, Siberia.* Silica 66.45, alumina 16.75, glucina 15.50, oxide of iron 0.60. *Klaproth.*—3. *Beryl, Broddbo.* Silica 68.35, alumina 17.60, glucina 13.13, oxide of iron 0.72, oxide of tantalum 0.27. *Berzelius.*

Geognostic and Geographic Situations.—The precious emerald is found in mica-slate and clay-slate; the ancients obtained their emeralds from Mount Zalora in Upper Egypt, but at present the finest varieties are those imported from Peru. The beryl, or common emerald, occurs in Aberdeenshire, in the Mourne mountains in Ireland, the United States, &c.; but nearly all the varieties met with in trade are brought from Russia.

Genus VI.—Quartz.

Rhombohedral prismatic. Cleavage not axotomous. Hardness = 5.5, — 7.5. Sp. gr. = 1.9, — 2.7.

1. *Prismatic Quartz, or Iolite.*

Prismatischer Quarz, *Mohs.*—Iolith. Peliom, *Werner.* —Iolithe, *Haüy.*—Dichroite. Cordierite. Steinheilite, *Auct.*

Specific Character.—Prismatic. $P + \infty = 120°$ nearly. Cleavage, $P + \infty$, $\check{P}r + \infty$. (Fig. 30, 29.) Hardness = 7.0, — 7.5. Sp. gr. = 2.5, — 2.6.

Description.—Colour blue. Of an indigo blue colour, when viewed in the direction of the axis, but viewed perpendicular to the axis, is brownish-yellow. Occurs rarely crystallized, generally massive and disseminated. Lustre shining and vitreous. Fracture uneven and conchoidal. Translucent in the direction of the axis, and transparent at right angles to it.

Constituent Parts.—Silica 48.54, alumina 31.73, magnesia 11.30, oxide of iron 5.69, oxide of manganese 0.70, water or loss 1.65. *Stromeyer.*

Geognostic and Geographic Situations.—Occurs in granite and gneiss; Abo in Finland, Arendal in Norway, St Gothard in Switzerland, and Granatillo in Spain, Ujondlersok in Greenland, are localities of this mineral.

2. *Rhombohedral Quartz.*

Quartz, *Werner.*—Rhomboedrisher Quarz, *Mohs.*

Specific Character.—Rhombohedral. Rhombohedron = 75° 55′. $P = 133° 44′$; 103° 35′. $\frac{P}{2} = 94° 15′$. Combination, hemirhombohedral and hemidirhombohedral. $R + n$ and $(P + n')^m$ with inclined planes, and $P + n''$ with parallel planes. Cleavage, P. $P + \infty$ imperfect. One-half of P generally still less perfect. Hardness = 7.0. Sp. gr. = 2.5,—2.7.

Description.—Colours white, grey, less frequently black, blue, green, yellow, red, and brown. Occurs regularly crystallized; in granular, fibrous, prismatic, and lamellar concretions; also massive, disseminated, in plates, stalactitic, reniform, botryoidal, globular, specular, vesicular, and cellular. Lustre ranges from splendent to glimmering, and is vitreous. Fracture conchoidal, even, uneven, and splintery. Ranges from transparent to opaque. Brittle, and easily frangible.

The transparent and highly crystallized varieties are named *rock-crystal.* The pyramidal, translucent, and transparent varieties, having generally a violet-blue colour, and disposed in prismatic, fibrous, and lamellar concretions, are named *amethyst.* The massive, strongly translucent, rose-red, and milk-white varieties, are named *rose-quartz.* Those varieties which have generally grey or white colours, the pyramidal form, and occur massive, disseminated, and in the various particular external forms enumerated above, with a low degree of translucency, are named *common quartz.* The leek-green translucent varieties, with a resino-vitreous lustre, and conchoido-splintery fracture, are named *prase.* The *cat's eye* is a variety, with a beautiful opalescence, like the light of the eye of the cat. Those opaque varieties, in which the colours are red, brown, and yellow, and the lustre shining externally, and glistening internally, and vitreo-resinous, and fracture small conchoidal, are named *iron-flint.* Those varieties which exhibit grey, yellow, brown, red, and green, generally muddy colours, which occur massive, and in extraneous external shapes, with a splintery or conchoidal fracture, with dull or glimmering lustre, and opaque or translucent on the edges, are named *hornstone.*

The grey, brown, and black varieties which generally occur massive, and in various particular and extraneous external forms, and exhibit a glimmering lustre, conchoidal fracture, and feeble translucency, are named *flint.* The semi-transparent and translucent varieties, in massive and various particular external forms, with an even and dull fracture surface, are named *common calcedony.* The beautiful apple-green strongly translucent varieties are named *chrysoprase.* The grass-green varieties, with glimmering lustre, conchoidal fracture, and strong translucency, are named *plasma.* The semi-transparent and strongly translucent varieties, with glistening or shining vitreous lustre and conchoidal fracture, and with various tints of red, brown, yellow, green, and white, are named *carnelian.* The beautiful variety, named *heliotrope*, is well characterized, on a general view, by its green colour, and disseminated spots of red and yellow jasper. Those varieties, in which the colours are red, brown, and black, and seldomer yellow or green, which occur massive and disseminated, with a fracture ranging from conchoidal to earthy, and lustre from glistening to dull, and are opaque, are named *jasper.* And, lastly, the white and grey varieties, which are so porous and light, as to swim on the surface of water, are named *spongiform quartz*, or *floatstone.*

Constituent Parts.—1. *Rock-crystal.* Silica 99.37, trace of alumina and oxide of iron. *Bucholz.*—2. *Amethyst.* Silica 97.50, alumina 0.25, oxide of iron a trace, manganese 0.75. *Rose.*—3. *Common Quartz.* Silica 97.75, alumina 0.50, water 1.00. *Bucholz.*—4. *Red Iron-flint.* Silica 76.8, alumina 0.25, oxide of iron 21.66, volatile matter 1.00. *Bucholz.*—5. *Carnelian.* Silica 94.00, alumina 3.50, oxide of iron 0.75. *Bindheim.*—6. *Chrysoprase.* Silica 96.16, alumina 0.08, lime 0.82. *Klaproth.*—7. *Common Flinty Slate.* Silica 51.84, alumina 15.43, lime 9.42, oxide of iron 9.96. *Dumeril.*

Geognostic and Geographic Situations.—Quartz is very universally distributed, and, as far as we know at present, is the most abundant mineral in nature. It occurs in every rock, from granite to the newest secondary formation; and every country, every district in the world, afford examples of this mineral.

3. *Uncleavable Quartz or Opal.*

Untheilbarer Quarz, *Mohs.*

Specific Character.—Reniform. Massive. No cleavage. Hardness = 5.5,—6.5. Sp. gr. = 1.9,—2.2.

Description.—Colours white and grey, also green, yellow, brown, red, and black, and sometimes displays a beautiful play of colour. Occurs massive, and in various particular external forms. Lustre ranges from splendent to glistening, and is vitreous, or vitreo-resinous. Ranges from transparent to opaque.

The grey and white varieties, with glistening and shining pearly lustre, and translucent on the edges, or translucent in the mass, are named *quartzy sinter*, or *pearl-sinter*. The yellowish and greyish white varieties with the botryoidal and other particular external forms, conchoidal fracture, and vitreous lustre, and which are strongly translucent, are named *hyalite*. The milk-white varieties, with the beautiful play of various rich and pure colours, are named *precious opal*. The transparent varieties, with the beautiful carmine-red and apple-green iridescence, are named *fire-opal*. Those varieties which are milk-white, and frequently dendritic, with a pearly, shining, or glistening lustre, conchoidal fracture, and complete opacity, are named *mother-of-pearl opal*, or *cacholong*. The *common opal* differs from the precious, principally in wanting the play of colour. The feebly translucent varieties, with conchoidal fracture, and glistening vitreo-resinous lustre, and white, grey, and brown colours, are named *semi-opal*. Those varieties, in which the colours are red, yellow, and grey, with a shining vitreo-resinous lustre, conchoidal fracture, and opacity, are named *jasper-opal*. Those varieties which occur in various vegetable forms, and are really vegetables impregnated, or petrified with opal, are named *wood-opal*. And lastly, the brown and yellowish grey tuberose varieties are named *menilite*.

Constituent parts.——1. *Precious Opal*. Silica 90.0, water 10.0. *Klaproth*.—2. *Fire-Opal from Mexico*. Silica 92.60, water 7.75, oxide of iron 0.25. *Klaproth*.—3. *Common Opal, Hungary*. Silica 93.5, water 5.0, oxide of iron 1.0. *Klaproth*.—4. *Semi-Opal*. Silica 82.75, water 10.00, oxide of iron 3.00, alumina 3.50. *Stucke*.—5. *Hyalite*. Silica 92.00, water 6.33. *Bucholz*.—6. *Menilite*. Silica 85.5, water 10.0.

Geognostic and Geographic Situations.—This species has a comparatively limited geognostic and geographic distribution. The quartzy sinter occurs in the vicinity of hot-springs, as in Iceland and other countries; the hyalite in secondary trap-rocks in Scotland, Germany; and the various opals are found principally in secondary trap, and in porphyry rocks in Hungary, Germany, Scotland and other countries.

4. *Fusible Quartz, Obsidian, Pitchstone, Pearlstone, Pumice.*

Empyrodox Quarz, *Mohs.*

Specific Character.—In grains...Massive. No cleavage. Hardness = 6.0,—7.0. Sp. gr. = 2.2,—2.4.

Description.—Colours black, green, and grey; also brown, blue, red, yellow, and white. Occurs in globular, lamellar, and prismatic concretions; massive, vesicular, and porous. Lustre ranges from splendent to glimmering, and is vitreous, resinous, and pearly. Ranges from transparent to opaque.

The varieties, with splendent vitreous lustre, conchoidal fracture, and ranging from transparent to translucent on the edges, are named *obsidian*, and *marekanite;* those again in which the lustre is shining and resinous, and are translucent on the edges, are named *pitchstone;* the beautiful varieties in globular and concentric lamellar, shining, pearly, translucent concretions, are named *pearlstone;* and lastly, the white and grey varieties, which are vesicular and fibrous, with a vitreous or pearly lustre, and often so light as to swim on water, are named *pumice.*

Constituent Parts.—1. *Obsidian, from Mexico.* Silica 78.0, alumina 10.0, lime 1.0, potash 6.0, oxide of iron 3.6. *Vauquelin.*—2. *Marekanite.* Silica 81.00, alumina 9.50, lime 0.33, natron 4.50, potash 2.70, water and oxide of iron 1.00. *Klaproth.*—3. *Pitchstone.* Silica 72.80, alumina 11.50, natron 2.85, lime 1.20, oxide of iron 3.03, water and bituminous matter 0.50.—4. *Pearlstone.* Silica 77.0, alumina 13.0, potash 1.5, lime and natron 2.7, oxide of iron and manganese 2.0, water 4.0. *Vauquelin.*—5. *Pumice.* Silica 77.50, alumina 17.50, natron and potash 3.00, oxide of iron with manganese 1.75. *Klaproth*

Geognostic and Geographic Situations.—Pitchstone is found on the mainland of Scotland as in Dumfriesshire; in the Islands of Arran, Egg, Rum, and Skye; pearlstone occurs in great beauty in Hungary and Iceland, and both pumice and obsidian are productions of Iceland, the Lipari Islands, and Mexico.

Genus VII.—Axinite.

Cleavage imperfect. Colour not inclining to yellow. Perfect vitreous lustre. Hardness = 6.5,—7.0. Specific gravity = 3.0,—3.3.

1. *Prismatic Axinite.*

Prismatischer Axinit, *Mohs.*—Thumerstein, *Werner.*—xinite, *Haüy.*

Specific Character.—Prismatic. Pyramid unknown. Combination tetarto-prismatic. Cleavage, two faces, the one more distinct than the other. Incidence about 101° 30′.

Description.—Colours brown, blue, and sometimes grey and black. Occurs regularly crystallized; in curved lamellar concretions, and massive. Lustre externally splendent, internally glistening or shining, and vitreous. Ranges from transparent to feebly translucent. Brittle, and very easily frangible.

Constituent Parts.—Silica 45.00, lime 12.75, alumina 10.00, oxide of iron 12.25, oxide of manganese 9.00, boracic acid 2.00. *Wiegmann.*

Geognostic and Geographic Situations. Occurs in beds and veins in gneiss, mica-slate, clay-slate, and hornblende-slate. It occurs in small quantities in Cornwall, but no where else in Great Britain.

Genus VIII.—CHRYSOLITE.

Prismatic. Colour green, brown, both inclining to yellow, yellow. Pure vitreous lustre. H = 6.5,—7.0. Sp. gr. = 3.15,—3.5.

1. *Prismatic Chrysolite.*

Prismatischer Chrysolithe, *Mohs.*—Peridot, *Haüy.*—Krisolith and Olivin, *Werner.*

Specific Character.—Prismatic. $\breve{P}r = 80° 53'$, $(\bar{P} + \infty)^2 = 130°2$. Cleavage $Pr + \infty$. Traces in the direction $Pr + \infty$. Colour green. Hardness = 6.5,—7.0. Sp. gr. = 3.3,—3.5.

Description.—Colour green. Occurs regularly crystallized. Lustre splendent and vitreous. Fracture conchoidal. Transparent.

The varieties in granular distinct concretions, which have a lower degree of transparency and lustre than the more highly crystallized kinds, are those generally described under the name *olivine.*

Constituent Parts.—1. *Chrysolite.* Silica 39.0, magnesia 43.5, oxide of iron 19.0. *Klaproth.*—2. *Olivine from Unkel.* Silica 50.0, magnesia 38.5, oxide of iron 12.0. *Klaproth.*—3. *Olivine from Meteoric Iron.* Silica 38.18, magnesia 48.82, oxide of iron 11.19. *Stromeyer.*—Traces of Nickel were observed.

Geognostic and Geographic Situations.—This mineral occurs principally in secondary trap-rocks. The olivine variety is met with in some of the middle and western districts of Scotland. The chrysolites of commerce are brought from the Brazils and Upper Egypt.

2. *Hemiprismatic Chrysolite or Chondrodite.*

Brucite. Maclureite.

Specific Character.—Hemiprismatic. Cleavage scarcely visible. Colours yellow, brown. Hardness = 6.5. Sp. gr. = 3.15,—3.25.

Description.—Colours straw, ochre, honey, orange, passing into hyacinth-red and yellowish-brown. Occurs in imbedded grains in small massive pieces, and in longish granular distinct concretions. Lustre vitreous. Fracture imperfect conchoidal. Translucent.

Constituent Parts.—Magnesia 54.00, silica 32.66, fluoric acid 4.08, oxide of iron 2.33, potash 2.10, water 1.00. *Seybert.*

Geognostic and Geographic Situations.—Occurs in calc spar and granular limestone in the parish of Pargas in Finnland; also in granular limestone, associated with graphite, at Newton in New Jersey, United States; at Lochness, and in granular limestone in Kincardineshire.

Genus IX.—Boracite.

Tessular. Hardness = 7. Sp. gr. = 2.8,—3.0.

1. *Octahedral Boracite.*

Octaedrischer Borazit, *Mohs.*—Borazit, *Werner.*—Magnesie Boratée, *Haüy.*

Specific Character.—Tessular. Combination semitessular of inclined planes. Cleavage octahedral but imperfect.

Description.—Colours white and grey. Occurs in cubes and dodecahedrons. Internally shining, and nearly adamantine. Fracture conchoidal. Translucent and transparent.

Constituent Parts.—1. Boracic acid 54.55, magnesia 30.68, oxide of iron 0.57, silica 2.27. *Pffaff.*—2. Boracic acid 69.95, magnesia 30.04.—*Arfwedson.*

Geognostic and Geographic Situations.—Occurs imbedded in gypsum, in the Kalkberg, in Hanover, and in the Segeberg, near Kiel, in Holstein.

Genus X.—Tourmaline.

Rhombohedral. Hardness = 7.0,—7.5. Sp. gr. = 3.0—3.2.

1. *Rhombohedral Tourmaline.*

Rhomboedrischer Turmaline, *Mohs.*—Tourmaline, *Haüy.*

Specific Character.—Rhombohedral.—Rhombohedron = 133° 26′. Combination hemirhombohedral, the opposite ends differently formed. Cleavage, R. $P + \infty$, imperfect.

Description.—Colours black, brown, green, blue, red, and white. Occurs regularly crystallized; in prismatic and granular distinct concretions, massive and disseminated. Lustre ranges from splendent to glistening, and is vitreous. Fracture conchoidal and uneven. Ranges from transparent to opaque. Brittle and easily frangible.

The black coloured opaque varieties, with uneven fracture, are named *schorl.*; all the others are ranged under the title *tourmaline.*

Constituent Parts.—1. *Black or Schorl of Rabenstein.* Silica 34.48, alumina 34.75, soda 1.75, potash 0.48, oxide of iron 17.44, oxide of manganese 1.89, *boracic acid* 4.02, magnesia 4.68. *Gmelin.*—2. *Blue Tourmaline, Uton.* Silica 40.30, alumina 40.50, oxide of iron 4.85, oxide of manganese 1.50, *boracic acid* 1.10, *lithion* 4.39. *Arfwedson.*—

3. *Green Tourmaline, Brazil.* Silica 39.16, alumina 40.00, oxide of iron 5.96, oxide of manganese 2.14, *boracic acid* 4.59, *lithion* 3.59. *Gmelin.*—4. *Red Tourmaline, Siberia.* Silica 39.37, alumina 44.00, potash 1.29, oxide of manganese 5.02, *boracic acid* 4.18, *lithion* 2.62. *Gmelin.*

Geognostic and Geographic Situations.—It occurs imbedded in granite, gneiss, mica-slate, talc-slate, chlorite-slate, dolomite, topaz-rock, and quartz-rock; also in felspar, mica, talc, &c. and in alluvial strata. The schorl varieties occur in various primitive rocks in the alpine districts of Scotland, and the purer varieties, or tourmaline, properly so called, are met with in Siberia, Spain, and many other countries.

Genus XI. Garnet.

Tessular, pyramidal, prismatic. Lustre not pure vitreous. Hardness = 6.0,—7.5. Sp. gr. = 3.1,—4.3. If colour black, sp. gr. = 3.9, and less. If hardness 7.5, the colour is red or brown. If sp. gr. less than 3.3, is tessular.

1. *Pyramidal Garnet or Vesuvian.*

Pyramidaler Granat, *Mohs.*—Vesuvian and Egeran, *Werner.*—Idocrase, *Haüy.*

Specific Character.—Pyramidal. Pyramid = 129° 29′; 74° 14′. Cleavage, P — ∞. P + ∞. [P + ∞]. Hardness = 6.5. Sp. gr. = 3.3,—3.4.

Description.—Colours green and brown, and rarely blue. Occurs regularly crystallized; in granular distinct concretions; massive, and disseminated. Externally splendent, and internally feebly shining and vitreo-resinous. Fracture uneven, inclining to small conchoidal. Ranges from transparent to translucent in the edges.

Constituent Parts.—Silica 38.85, alumina 21.93, lime 33.61, oxide of iron 5.40. *Kobell.*

Geognostic and Geographic Situations.—Occurs in primitive rocks in Ireland, also in Scotland; and on the continent of Europe the most beautiful varieties are found in the ejected masses on Somma, near Naples, where they are associated with granular limestone, garnet, hornblende, mica, chlorite, augite, meionite, nepheline, magnetic iron-ore, &c.

2. *Tetrahedral Garnet, or Helvine.*

Helvin, *Werner.*—Tetraedrischer Granat, *Mohs.*

Specific Character.—Tessular. Combination semi-tessular of inclined planes. Cleavage octahedral, but imperfect. Hardness = 6.0,—6.5. Sp. gr. = 3.1,—3.3.

Description.—Colours wax-yellow, inclining to yellowish-brown or siskin-green. Occurs regularly crystallized, in tetrahedrons, truncated on the angles, and in granular distinct concretions. Lustre ranges from splendent to glistening and vitreo-resinous. Fracture uneven. Translucent, or translucent on the edges. Streak white.

Constituent Parts.—Silica 35.27, alumina 1.45, oxide of iron 7.99, oxide of manganese 29.44, sulphate of manganese 14.00, glucina 8.03. *Gmelin.*

Geognostic and Geographic Situations.—Occurs in a bed in gneiss, associated with fluor-spar, slate-spar, chlorite, quartz, blende, and copper-pyrites, near Schwarzenberg, in Saxony, also in brown ironstone, near Breïtenbrun, also in Saxony.

3. *Dodecahedral Garnet.*

Specific Character.—Tessular. Cleavage dodecahedral, imperfect. Hardness = 6.5,—7.5. Sp. gr. = 3.5,—4.3

Description.—Colours red, brown, green, black, grey, and yellow. Occurs regularly crystallized; in angulo-granular concretions, and massive. Lustre ranges from splendent to glimmering, and is vitreous inclining to resinous, resino-vitreous, and resino-adamantine. Fracture conchoidal, splintery, uneven. Ranges from transparent to opaque.

The asparagus-green varieties in leucite formed crystals are named *grossulare;* the greyish-black dodecahedral opaque varieties, which externally have a metallic-like aspect, *pyrenaite;* the velvet-black, dodecahedral opaque varieties, *melanite;* the green, brown, and grey massive varieties, with glimmering lustre, and feeble translucency on the edges, *allochroite;* the brown and red varieties in granular concretions, with resino-adamantine lustre, *colophonite;* the red highly crystallized transparent varieties, *precious garnet;* the yellow nearly transparent varieties, *topazolite* or *yellow garnet;* the brown and green varieties, in crystals often rounded on the edges, in granular concretions, and translucent, or translucent on the edges, *common garnet;* the deep blood-red variety in roundish and angular grains, and completely transparent, *pyrope;* and the hyacinth and orange-yellow varieties, in granular distinct concretions, and massive, with resino-vitreous lustre, conchoidal fracture, and transparent or translucent, are named *Essonite* or *Cinnamon Stone.*

Constituent Parts.—1. *Grossulare.* Silica 40.55, alumina 20.10, lime 34.86, oxide of iron 5.00, oxide of manganese 0.48. *Wachtmeister.*—2. *Pyrenaite.* Silica 43.0, alumina 16.0, lime 20.0, oxide of iron 16.0, water 4.0. *Vauquelin.*—3. *Melanite.* Silica 35.5, alumina 6.0, lime 32.5, oxide of iron 25.2, oxide of manganese 0.4.—4. *Al-*

Iochroite. Silica 35.0, alumina 8.0, lime 30.0, oxide of iron 17.0, oxide of manganese 3.5. *Vauquelin.*—5. *Colophonite.* Silica 37.0, alumina 13.5, lime 29.0, oxide of iron 7.5, oxide of manganese 4.8, magnesia 6.5, water 1.0. *Simon.*—6. *Precious Garnet.* Silica 35.75, alumina 27.25, oxide of iron, 36.00, oxide of manganese 0.25.—*Klaproth.*—7. *Pyrope.* Silica 40.0, alumina 28.5, lime 3.5, oxide of iron 16.5, oxide of manganese 0.3, magnesia 10.0, chromic acid 2.6.—8. *Essonite or Cinnamon Stone.* Silica 38.80, alumina 21.26, lime 31.25, oxide of iron 6.50. *Klaproth.*

Geognostic and Geographic Situations.—This mineral occurs principally in primitive rocks, either disseminated, or forming an ingredient in the composition of subordinate beds; it is comparatively rare in transition rocks, and is still less frequently met with in secondary formations. In Scotland the *precious garnet* is common in several Highland districts in Perthshire, Aberdeenshire, &c.; the *pyrope* variety occurs in serpentine rocks in Bohemia; the *cinnamon stone* or *essonite* in gneiss in Kincardineshire; the *melanite* or black garnet is found in Vesuvius, and in the neighbourhood of Rome; the *grossulare* is met with near the river Willui in Siberia; the *pyrenaite* in the Pic Eres Lids, near Barèges, in the Pyrenees; the *colophonite* chiefly at Arendal, in Norway; *allochroite* in the neighbourhood of Drammen, in Norway; and the finest *cinnamon stones* are the produce of Ceylon.

4. *Prismatoidal Garnet, or Grenatite.*

Prismatoidischer granat, *Mohs.*—Granatit, *Werner.*—Staurotide, *Haüy.*

Specific Character.—Prismatic. $\bar{P}r = 70° \ 32'$;

$(\bar{P} + \infty)^2$ = 129° 31′. Cleavage, $\breve{P}r + \infty$, perfect. Traces in direction $(\bar{P} + \infty)^2$. (Fig. 29.) Hardness = 7.0,—7.5. Sp. gr. = 3.3,—3.9.

Description.—Colour reddish-brown. Occurs regularly crystallized. Internally the cleavage is shining and splendent; fracture glistening and glimmering, and resino-vitreous. Fracture uneven. Ranges from semi-transparent to opaque. Streak white.

Constituent Parts.—Silica 33.00, alumina 45.00, lime 3.84, oxide of iron 13.00, oxide of manganese 4.00. *Vauquelin.*

Geognostic and Geographic Situations.—Occurs in primitive rocks in the Shetland Islands, in Aberdeenshire, and other parts of Scotland.

Genus XII.—Zircon.

Pyramidal. Hardness = 7.5. Sp. gr. = 4.5,—4.7.

1. *Pyramidal Zircon.*

Pyramidaler Zircon, *Mohs.*

Specific Character.—Pyramidal. Pyramid = 123° 19′; 84° 20′. Cleavage, P. $P + \infty$.

Description.—Colours grey and hyacinth red; also white, green, brown, and rarely yellow, blue, and red. Occurs regularly crystallized. Lustre splendent, shining, and adamantine-resinous and resino-vitreous. Fracture conchoidal. Ranges from transparent to opaque.

Those varieties in which the cleavage is very distinct, and which have frequently a hyacinth-red colour, are named *hyacinth*, the other varieties *common zircon.*

Constituent Parts.—1. *Zircon.* Zirconia 69.0, silica 26.5, oxide of iron 0.5. *Klaproth.*—2. *Hyacinth.* Zirconia 70.0, silica 25.0, oxide of iron 0.5. *Klaproth.*

Geognostic and Geographic Situations.—Occurs in syenite, granite, gneiss, primitive trap, in secondary trap, and ejected masses of Somma. The syenite rocks of Galloway, the gneiss rocks of Inverness-shire, and of the Shetland Islands, afford examples of this mineral. The most esteemed varieties of zircon, are from the Island of Ceylon.

Order VIII.—ORE.

Streak not green. Hardness = 2.0,—7,0. Sp. gr. = 3.4,—7.4. If metallic, the colour is black. If not metallic, the lustre is adamantine or imperfect metallic. If streak yellow or red, the hardness equal to 3.5 and more; and sp. gr. 4.8 or more. If streak brown or black, the hardness is equal to 5.0 or more, or 2.5 and less; or single distinct cleavage. If hardness is 2.5 or less, the specific gravity is under 5.0. If hardness is 4.5 or less, the streak is yellow, red, or black. If hardness equal to 6.5 or more, and streak uncoloured, the specific gravity equal 6.5 and more.

Genus I.—Melane-Ore.

Prismatic. Black. Streak greenish and brownish-grey. Hardness = 5.5,—7.5. Sp. gr. = 3.2,—4.3.

1. *Prismatic Melane-Ore or Gadolinite.*

Gadolinite.—*Mohs*, *Haüy*, and *Haidinger.*

Specific Character.—Indistinct crystals which are prisms of about 109° 28′. Colour greenish-black. Streak greenish-grey. Hardness = 6.5 —7.0. Sp. gr. = 4.2,—4.3.

Description.—Colours black, or greenish-black. Occurs

massive; in granular and prismatic concretions. Internally shining, and lustre resino-vitreous. Fracture conchoidal, seldom uneven. Opaque, or very faintly translucent on the edges.

Constituent Parts.—Yttria 45.93, protoxide of iron 11.34, protoxide of cerium 16.90, silica 24.16. *Berzelius.*

Geognostic and Geographic Situations.—Occurs in beds of felspar, or mica-slate at Ytterby, and in granite at Finbo, in Sweden; also in Disco Island in Davis' Straits, and in granite in the Island of Ceylon.

2. *Anorthitic Melane-Ore or Allanite.*

Cerin, *Hisinger and Berzelius.*—Cerium Oxydé Siliceux, *Haüy.*—Allanite, *Thomson.*—Anorthisches Melan-Erz including Orthite, *Haidinger.*

Specific Character.—Tetarto-prismatic. Prism = 115°—deeply truncated on the obtuse edges. Cleavage imperfect. Greenish and brownish-black. Streak greenish and brownish-grey. Hardness = 5.5,—6.0. Sp. gr. = 3.2,—3.6.

Description.—Colours black, inclining to green, grey, or brown. Occurs massive or in acicular crystals. Lustre imperfect, metallic. Fracture conchoidal. Opaque. Streak greenish or brownish-grey.

Constituent Parts.—1. *Allanitic Melane-Ore.* Oxide of cerium 33.9, oxide of iron 25.4, silica 35.4, lime 9.2, alumina 4.1, water 4.0. *Thomson.*—2. *Orthitic Melane-Ore.* Silica 32.00, lime 7.84, alumina 14.80, oxide of cerium 19.44, protoxide of iron 12.44, oxide of manganese 3.40, yttria 3.44, water 5.36. *Berzelius.*

Geognostic and Geographic Situations.—The Allanite varieties occur at Alluk, near the southern extremity of Old Greenland. The varieties named Orhite occur in quartz at Finbo, near Fahlun, in Sweden, and in granite at Skeppsholm, one of the islands on which Stockholm is built, also at Lindenaes, in Norway, and in Greenland. It was named Allanite in honour of the late Mr Thomas Allan, one of our best mineralogists, and distinguished as an accurate and judicious geologist.

Genus II.—Titanium-Ore.

Pyramidal, prismatic. Streak white, or very pale brown. Hardness = 5.0,—6.5. Sp. gr. = 3.4,—4.4. If the sp. gr. less than 4.2, the streak is white.

1. *Hemiprismatic Titanium-Ore, or Sphene.*

Prismatisches Titan-erz, *Mohs.*—Gelb and Braun, Manak-erz, *Werner.*—Titane Siliceo-calcaire, *Haüy.*—Hemiprismatisches Titan-erz, *Haidinger.*

Specific Character.—Prismatic. Hemiprismatic $\frac{P}{2}$ = 113° 20′. Inclination = 8° 18′. $(Pr + \infty)^2 = 136° 8'$. Cleavage, $\frac{P}{2} \cdot \frac{Pr}{2} = 28° 7'$, difficult. (Fig. 37). Streak white. Hardness = 5.0,—5.5. Sp. gr. = 3.4,—3.6.

Description.—Colours brown, yellow, green, grey, and white. Occurs regularly crystallized; in granular and lamellar distinct concretions, and massive. Lustre ranges from splendent to glistening, and is adamantine. Fracture conchoidal. Ranges from transparent to opaque.

One set of varieties, in which brown is the predominating colour, is named *common sphene*, or *brown titanium-ore;* and another, in which the principal colours are yellow, and the cleavage distinct, is named *yellow titanium-ore*, or *foliated sphene.*

Constituent Parts.—Oxide of titanium 33.0, silica 35.9, lime 33.0, trace of manganese. *Klaproth.*

Geognostic and Geographic Situations.—This mineral occurs imbedded in syenite in Inverness-shire, Perthshire,

Galloway, and many other quarters of Scotland ; it also occurs in syenite rocks in England.

2. *Octahedral Titanium-Ore, or Pyrochlore.*

Octaedrisches Titan-erz, *Mohs.*—Pyrochlor, *Wöhler.*

Specific Character.—Tessular. Cleavage scarcely perceptible. Streak pale brown. Hardness = 5.0. Sp. gr. = 4.2,—4.25.

Description.—Colour dark reddish-brown, passing into blackish-brown. Streak pale brown. Occurs massive, disseminated, and in small octahedrons. Lustre shining and vitreo-resinous. Fracture conchoidal. From translucent on the edges to opaque.

Constituent Parts.—Titanic acid 62.75, lime 12.85, protoxide of uranium 5.18, oxide of cerium 6.80, protoxide of manganese 2.75, oxide of tin 0.61, oxide of iron 2.16, water 4.20. *Wöhler.*

Geognostic and Geographic Situations.—Occurs imbedded in syenite at Laurvig and Fredereickswärn, Norway; also in Greenland.

3. *Peritomous Titanium-Ore, or Rutile.*

Peritomes Titan-erz, *Mohs.*—Titan-oxydé, *Haüy.*—Rutil. Nigrine, *Werner.*—Titan schorl. Sagenite.

Specific Character.—Pyramidal. Pyramid = 117° 2′; 95° 13′. Cleavage, P + ∞. [P + ∞]. Streak pale brown. Hardness = 6.0,—6.5. Sp. gr. =4.2,—4.4.

Description.—Colours brown, red, yellow, and sometimes nearly velvet black. Occurs regularly crystallized, massive, disseminated, in angular grains and in flakes. Lustre metallo-adamantine and adamantine, and ranges from splenpent to glistening. Fracture conchoidal. Ranges from transparent to opaque.

The dark brown and black opaque varieties are named *nigrine;* the others *rutile.*

Constituent Parts.—The pure varieties contain of titanium 56.05, and oxygen 33.35; the black varieties, named nigrine, contain oxide of titanium 84.0, oxide of iron 14.0, oxide of manganese 2.0.

Geognostic and Geographic Situations.—The rutile varieties occur in the granite of Cairngorm; at Killin, and in Ben-Gloe, in quartz-rock. The nigrine varieties are met with at Ely, in Fifeshire, and also in Bohemia and Transylvania.

4. *Pyramidal Titanium or Anatase.*

Pyramidales Titan-erz, *Mohs.*—Octaedrit, *Werner.*—Titane Anatase, *Haüy.*—Oisanite, *Lameth.*

Specific Character.—Pyramidal. Pyramid = 70° 56′; 136° 22′. Cleavage, P — ∞, P. Perfect. Streak white. Hardness = 5.5,—6.0. Sp. gr. = 3.8,—3.9.

Description.—Colours blue and brown. Occurs regularly crystallized. Lustre splendent and adamantine, inclining to semi-metallic. Translucent and transparent.

Constituent Parts.—Is a pure oxide of titanium.

Geognostic and Geographic Situations.—Occurs in small veins in primitive rocks, as greenstone, gneiss, mica-slate, clay-slate, granite, in Dauphiny, Spain, Norway, and Brazil.

5. *Prismatic Titanium-Ore, or Brookite.*

Prismatisches Titan-erz, *Haidinger.*—Brookite, *Levy.*

Specific Character.—Prismatic. Prism = 100° and 80°. Colour hair-brown. Streak yellowish-white. Hardness = 5.5,—6.0. Sp. gr. unknown.

Description.—Colour hair-brown, inclining to dark

orange-yellow and red. Streak yellowish-white. Lustre metallic, adamantine. Translucent or opaque. Brittle.

Constituent Parts.—Chiefly composed of oxide of titanium.

Geognostic and Geographic Situations.—Occurs along with quartz and anatase at Bourg d'Oisans in Dauphiny, the Tete-Noire in Savoy, and in the mountain of Snowdon in Wales.

Genus III.—Zinc-Ore. Red Oxide of Zinc.

Prismatic. Streak orange-yellow. Hardness =4.0,—4.5. Sp. gr. = 5.4,—5.5.

1. *Prismatic Zinc-Ore.*

Prismatisches Zink-erz, *Mohs* and *Haidinger.*—Zinc Oxide Ferrifere Lamellaire Brun-Rougeatre, *Haüy.*

Specific Character.—Prismatic. Pr = 125° 25′, nearly. The principal cleavage parallel to the terminal plane of the prism.

Description.—Colour between blood-red and aurora-red. Streak orange-yellow. Massive, disseminated, and in granular concretions. Lustre adamantine, and shining. Translucent on the edges. Fracture conchoidal. Brittle.

Constituent Parts.—Oxide of zinc 92.0, oxides of iron and manganese 8.0. *Bruce.*—Oxide of zinc 88.0, oxides of iron and manganese 12.0. *Berthier.*

Geognostic and Geographic Situations.—This important ore has hitherto been found only in America, where it occurs along with calcareous spar, and Franklinite at Sparta, in New Jersey.

Genus IV.—Red Copper-Ore.

Tessular. Streak brownish-red. Hardness = 3.5,—4.0. Sp. gr. = 5.6,—6.1.

1. *Octahedral Red Copper-Ore.*

Octaedrisches Kupfererz, *Mohs.*—Roth Kupfererz, *Werner.*—Cuivre oxydulé and Cuivre oxide rouge, *Haüy.*

Specific Character.—Tessular. Cleavage octahedral.

Description.—Colours red, particularly cochineal-red. Occurs crystallized, frequently in cubo-octahedrons; in granular distinct concretions; massive, disseminated, and in flakes. Lustre ranges from shining to glimmering, and is adamantine, inclining to semi-metallic. Fracture uneven. Ranges from translucent to opaque. Brittle.

The varieties with cleavage are named *foliated red copper-ore;* those which are massive, glistening, and opaque, *compact red copper-ore;* and the varieties in capillary crystals, *capillary red copper-ore.*

Constituent Parts.—Copper 88.5, oxygen 11.5. *Chenevix.*

Geognostic and Geographic Situations.—Occurs in veins in gneiss, mica-slate, clay-slate, and greywacke; also in veins and beds in secondary rocks. The copper mines of Cornwall afford fine examples of this beautiful ore.

Genus V.—Tin-Ore. Oxide of Tin.

Zinn-erz, *Mohs.*

Pyramidal. Streak not black. Hardness = 6.0,—7.0. Sp. gr. = 6.3,—7.1.

1. *Pyramidal Tin-Ore.*

Pyramidales Zinn-erz, *Mohs.*—Etain oxydé, *Haüy.*

Specific Character.—Pyramidal. Pyramid = 133° 26′; 67° 59′. Cleavage P + ∞. [P + ∞]. Streak white or pale brown.

Description.—Colours brown, black, green, white, yellow, and red. Occurs regularly crystallized, very generally in four-sided prisms acuminated with four planes, and often in twin crystals; also reniform, botryoidal, globular, and in delicate fibrous concretions. Externally splendent; internally ranges from splendent to glimmering, and is adamantine, inclining to resinous. Fracture uneven. Ranges from semi-transparent to opaque.

The varieties in fibrous concretions are named *wood-tin*, the others *common tinstone.*

Constituent Parts.—Oxide of tin 95.0, oxide of iron 5.0. *Descotils.*

Geognostic and Geographic Situations.—Occurs in granite, gneiss, mica-slate, clay-slate, porphyry, and in alluvial depositions. The mines of Cornwall afford all its varieties.

Genus VI.—Wolfram-Ore.

Scheel-erz, *Mohs.*

Prismatic. Streak dark reddish-brown. Hardness = 5.0, —5.5. Sp. gr. = 7.1,—7.4.

1. *Prismatic Wolfram. Tungstate of Iron.*

Prismatisches Scheel-erz, *Mohs.*—Wolfram, *Werner.*—Scheelin ferruginé, *Haüy.*

Specific Character.—Prismatic. Oblique prism = 105° 5′. Cleavage $\breve{P}r + \infty$, perfect. (Fig. 29.)

Description.—Colour brownish-black. Occurs regularly crystallized, and massive. Lustre ranges from splendent to glistening, and is adamantine, inclining to semi-metallic. Fracture uneven. Opaque.

Constituent Parts.—Tungstic acid 78.77, protoxide of manganese 6.22, protoxide of iron 18.32, silica 1.25. *Berzelius.*

Geognostic and Geographic Situations.—Occurs in primitive rocks in the island of Rona, one of the Hebrides, in Cornwall, in the Saxon and Bohemian tin-mines, &c.

Genus VII.—Tantalum-Ore.

Tantal-erz, *Mohs.*

Prismatic. Streak brownish-black. Hardness = 6.0. Sp. gr. 6.0,—6.3.

1. *Prismatic Tantalum-Ore.*

Prismatisches Tantal-erz, *Mohs.*—Tantale Oxidé Ferro-Manganesifere, *Haüy.*

Specific Character.—Prismatic. Cleavage prismatoidal.

Description.—Colour black. Occurs regularly crystallized; massive and disseminated. Lustre shining and glistening, and semi-metallic-adamantine. Fracture uneven, or conchoidal. Opaque.

Constituent Parts.—Oxide of tantalum 67.6, oxide of tin 8.7, oxide of iron 7.6, oxide of manganese 5.9, oxide of scheelium 8.7, lime 1.5. *Berzelius.*

Geognostic and Geographic Situations.—Occurs in granite in Finland; in granite, along with beryl, iolite, uran-mica, and iron-pyrites, at Bodenmais in Bavaria, also in North-America.

Observations.—Berzelius mentions a variety of this species with a pitch-black streak, and of rather higher specific gravity, which he thinks may prove a new species.

Genus VIII.—Fergusonite.

Fergusonite, *Haidinger.*

Pyramidal. Colour brownish-black. Streak pale brown. Hardness = 5.5,—6.0. Sp. gr. = 5.8,—5.9.

1. *Pyramidal Fergusonite.*

Specific Character.—Pyramidal. Terminal edge = 100° 28′, edge of base = 128° 27′.

Constituent Parts.—Columbic acid 47.75, yttria 41.91, oxide of cerium 4.68, zirconia 3.02, oxide of tin 1.00, oxide of uranium 0.95, oxide of iron 0.34. *Hartwall.*

Geognostic and Geographic Situations. This interesting mineral occurs in quartz, at Kikertausak, near Cape Farewell, in Greenland, where it was found by Gieseké.

Observation.—It is named in honour of an excellent mineralogist, Mr Ferguson of Raith.

Genus IX.—Uranium-Ore.

Uran-erz, *Mohs.*

Form unknown. Streak black. Hardness = 5.5. Sp. gr. = 6.4,—6.6.

1. *Uncleavable Uranium-Ore.*

Untheilbares Uran-erz, *Mohs.*—Uran-pecherz, *Werner.*—Uran oxydulé, *Haüy.*

Specific Character.—Uncleavable, reniform, and massive. No discernible cleavage.

Description.—Colour greyish-brown and velvet-black. Occurs massive, and reniform. Lustre shining, and adamantine, inclining to semi-metallic. Fracture conchoidal, passing into uneven. Opaque.

Constituent Parts.—Protoxide of uranium 86.5, protoxide of iron 2.5, silica 5.0, sulphuret of lead 6.0. *Klaproth.*

Geognostic and Geographic Situations.—Occurs in veins in primitive rocks, along with native silver, red silver, iron and copper pyrites, galena, blende, and brown-spar, in Saxony, Bohemia, and Hungary. The only British locality is Cornwall.

Genus X.—Cerium-Ore.

Cerer-erz, *Mohs.*

Rhombohedral. Streak white. Hardness = 5.5. Sp. gr. = 4.9,—5.0.

1. *Rhombohedral Cerium-Ore.*

Rhomboedrisches Cerer-erz, *Haidinger.*

Untheilbares Cerer-erz, *Mohs.*—Cerer-erz, *Werner.*—Cerium oxidé silicifère. *Haüy.*

Specific Character.—Rhombohedral. Low regular six-sided prisms. Massive. Traces of cleavage.

Description.—Colours red and brown. Occurs massive and disseminated. Internally glimmering and adamantine. Fracture splintery. Opaque.

Constituent Parts.—Oxide of cerium 68.59, silica 18.00, oxide of iron 2.00, lime 1.25, water and carbonic acid 9.60. *Hisinger.*

Geognostic and Geographic Situations.—Occurs in a bed of copper pyrites in gneiss, near Ridderhyttan, in Westmannland, in Sweden.

* *Cerine.*

Cerium Oxide Siliceux Noire, *Haüy.*

Colour brownish-black. Prismatic. Cleavage prismatoidal. Streak yellowish-grey. Hardness = 5.0.—60. Sp. gr. = 4.1,—4.2. Found in Sweden.

Constituent Parts.—Silica 30.17, alumina 11.31, lime 9.12, oxide of cerium 28.19, oxide of iron 20.72.

Geognostic and Geographic Situations.—Is found associated with rhombohedral cerium-ore at Bastnaes near Riddarhyttan in Sweden.

Genus XI.—Chrome-Ore.

Chrom-erz, *Mohs.*

Tessular. Streak brown. Hardness = 5.5. Sp. gr. = 4.4,—4.5.

1. *Octahedral Chrome-Ore, or Chromate of Iron.*

Octaedrisches Chrom-erz, *Mohs.*—Chrom-eisenstein, *Werner.*—Fer chromaté, *Haüy.*

Specific Character.—Tessular. Cleavage octahedral, but imperfect.

Description.—Colour iron-black or brownish-black. Occurs regularly crystallized; also in granular distinct concretions; massive and disseminated. Internally shining and glistening, and imperfect metallic. Fracture uneven. Opaque.

Constituent Parts.—Barehills, near Baltimore, United States. Green oxide of chrome 52.95, peroxide of iron 29.24, alumina 12.22, white matter 3.09. If the green oxide of chrome be considered as acting the part of an acid, this ore may be considered as composed of 1 atom chromite of iron, and 1 atom of chromite of alumina. *Thomson.*

Geognostic and Geographic Situations.—Occurs in imbedded masses, and in veins in serpentine, porphyry, and secondary trap. In Scotland it occurs principally in serpentine rocks, and more abundantly in the Shetland Islands of Unst and Ferlar, than elsewhere in this country.

Genus XII.—Iron-Ore.

Eisen-erz, *Mohs.*

Tessular, rhombohedral, prismatic. Streak red, brown, black. Hardness = 5.0,—6.5. Sp. gr. = 3.8,—5.3. If streak brown, the specific gravity = 4.2, and less, and form prismatic; or = 4.8, and more. If the streak is black, and the specific gravity = 4.8, and more, it acts on the magnet. If the specific gravity is under 4.3, and the colour black, the streak is not shining.

1. *Octahedral Iron-Ore.—Black or Magnetic Iron-Ore.*

Octaedrisches eisenerz, *Mohs.*—Fer oxydulé, *Haüy.*—Magnet-eisenstein, *Werner.*

Specific Character.—Tessular. Cleavage octahedral. Streak black. Hardness = 5.5,—6,5. Sp. gr. = 4.8,—5.2.

Description.—Colour iron-black. Occurs regularly crystallized; in granular distinct concretions; in loose grains; massive and disseminated. Lustre ranges from splendent to glistening, and is metallic. Fracture uneven, or conchoidal. Opaque. It is magnetic.

The variety in grains is named *iron-sand.* The other varieties, *common magnetic iron-ore.*

Constituent Parts.—Consists entirely of iron and oxygen, in the proportion of 2 atoms peroxide to 1 protoxide; contains 28.21 per cent. of oxygen.

Geognostic and Geographic Situations.—Occurs in beds, often of vast thickness and great extent, in rocks of the older formations, as gneiss, mica-slate, hornblende-slate, clay-slate, and primitive greenstone; also variously disposed in granite, syenite, serpentine, and chlorite-slate; less frequently in transition-rocks, in veins, beds, and imbedded masses, as in transition porphyry; and still less frequently in secondary trap-rocks. The serpentine, chlorite, and gneiss-rocks of the Shetland Islands afford examples of this ore; the same is the case on the Mainland of Scotland, both in primitive and secondary rocks. The great beds of iron-ore at Arendal in Norway, and the chief ore in nearly all the iron-mines in Sweden, is magnetic iron-ore.

2. *Rhombohedral Iron-Ore.—Specular Iron-Ore.—Iron-Glance.—Red Iron-Ore.—Red Hematite.*

Rhomboedrisches Eisen-erz, *Mohs.*—Fer oligiste, *Haüy.* Eisen-glanz. Roth Eisenstein, *Werner.*

Specific Character.—Rhombohedral. Rhombohedron $= 85° 58'$. Cleavage R, sometimes $R - 8$. Streak red,... reddish-brown. Sometimes feebly magnetic. Hardness $=$ 5.5,—6.5. Sp. gr. $= 4.8$,—5.3.

Description.—Colours dark steel-grey bordering on iron-black, iron-black, and brownish-red. Occurs regularly crystallized; in granular, lamellar, and fibrous distinct concretions; massive, disseminated, reniform, botryoidal, stalactitic, and globular. Lustre ranges from splendent to dull, and is metallic and semi-metallic. Fracture conchoidal or earthy. Ranges from translucent to opaque.

The dark steel-grey and iron-black varieties, which are generally more or less regularly crystallized, are named *specular iron-ore,* or *iron-glance;* the red varieties are named *red iron-ore,* and of these the fibrous varieties are named *hematite.*

Constituent Parts.—Specular iron-ore is a peroxide of iron, the proportion of iron 69.34, to oxygen 30.66. Red hematite affords of peroxide of iron 94.0, silica 2.0, lime 1.0, water 3.0. *Daubuisson.*

Geognostic and Geographic Situations.—The specular iron-ore occurs in gneiss, granite, mica-slate, transition clay-slate, greywacke, and less frequently in secondary rocks. The island of Elba affords the richest and most beautiful varieties of specular iron-ore, and specimens of considerable beauty are met with in Fitful-head in Shetland, and near Dunkeld in Perthshire. The red iron-ore occurs also in primitive rocks, but less frequently than in those of the transition class, as greywacke and transition clay-slate; and considerable deposits are met with in secondary limestone districts. It occurs abundantly near to Ulverstone in Lancashire, in smaller quantities in other parts of England, and also in Scotland.

3. *Prismatic Iron-Ore. Brown Iron-Ore. Brown Hematite.*

Prismatisches Eisen-erz, *Mohs.*—Braun Eisenstein, *Werner.*—Fer oxydé, *Haüy.*—Fer Hydraté, *Daubuisson.*

Specific Character.—Prismatic. $\bar{P}r = 117° 30'$. $P + \infty = 130° 14'$. Cleavage $= P + \infty$. Streak yellowish-brown. Does not affect the magnet. Hardness = 5.0,—5.5. Sp. gr. = 3.8,—4.2.

Description.—Colours brown and yellow. Occurs regularly crystallized; in granular, fibrous, and lamellar distinct concretions; massive, stalactitic, coralloidal, reniform, botryoidal, tuberose, cylindrical, and fruticose. Lustre glimmering and semi-metallic, inclining more or less to adamantine. Fracture uneven, even, conchoidal, or earthy. Translucent on the edges, or opaque; but the crystals often semitransparent, and showing a blood-red colour.

Constituent Parts.—*Hematitic or fibrous variety.* Peroxide of iron 82.0, water 14.0, oxide of manganese 2.0, silica 1.0, *Daubuisson.*—*Compact variety.* Peroxide of iron 84.0, water 11.0, oxide of manganese 2.0, silica 2.0, *Daubuisson.* It is thus a hydrous peroxide of iron; the proportion of peroxide of iron and water being as 85.30 to 14.70.

Geognostic and Geographic Situations.—Occurs in veins, beds, lenticular masses, and mountain masses, in primitive, transition, and secondary rocks. At Clifton, near Bristol, and near to Lostwithiel in Cornwall, beautiful specimens of the finest varieties are found, and the Shetland and Orkney

Islands, Cumberhead in Lanarkshire, are also localities of this ore.

Bog Iron-Ore.

Limonit. Fer Oxyde, des Lacs des Marais, &c.—Raseneisenstein, *Werner*.

Description.—Colour brown. Occurs massive, vesicular, corroded, amorphous, and tuberose. Some varieties are friable. Lustre ranges from glistening to dull, and is semi-metallic-resinous. Fracture earthy or conchoidal. Opaque, Yields a yellowish-grey streak. Brittle, and easily frangible. Sp. gr. from 2.6 to 3.0.

Constituent Parts.—Oxide of iron 66.0, phosphoric acid 8.0, water 23.0, oxide of manganese 1.5. *Klaproth.*

Geognostic and Geographic Situations.—It occurs in alluvial soil and in peat-mosses in various places in the mainland of Scotland, and also in the Orkney, Shetland and Western Islands.

4. *Diprismatic Iron-Ore, or Lievrite. Yenite. Ilvaite.*

Fer Calcareo-siliceux, *Haüy.*

Specific Character.—Pyramid $= 139° \ 37'$; $117° \ 38$: $77° \ 16'$. $P + \infty = 112° \ 37'$. Cleavage $\bar{P}r = 113° \ 2'$; $P + \infty$; $P - \infty$; $\mathring{P}r + \infty$. The whole, but particularly the first, imperfect. Streak black, sometimes inclining to green and brown. Does not affect the magnet. Hardness $= 5.5, - 6.0$. Sp. gr. $= 3.8, - 4.1$.

Description.—Colour black, blackish-green. Occurs regularly crystallized; also in distinct concretions, which are scopiform radiated, or straight radiated. Lustre glistening and semi-metallic. Fracture uneven. Opaque.

Constituent Parts.—Oxide of iron 52.54, silica 29.28, lime 13.78, oxide of manganese 1.58, water 1.27. *Stromeyer.*

Geognostic and Geographic Situations.—Occurs associated with epidote, garnet, magnetic iron-ore, and arsenic-pyrites, in limestone, in the island of Elba, and has also been found in Norway and West Greenland.

Genus XIII.—Manganese-Ore.

Mangan-erz, *Mohs.*

Pyramidal. Prismatic. Streak dark-brown, black. Does not affect the magnet. Hardness = 2.0,—6.5. Sp. gr. = 4.0,—4.9. If streak brown, the sp. gr. = 4.7, and more; or hardness = 4.0, and less. If hardness above 4.0; sp. gr. = 4.2, and less; and streak black: it is shining in the streak.

1. *Pyramidal Manganese-Ore, or Black Manganese-Ore. Hausmannite.*

Pyramidales Mangan-erz, *Mohs.*—Schwarzer Braunstein, *Werner.*—Manganese oxydé hydraté, *Haüy.*

Specific Character.—Pyramidal. Pyramid = 105° 25′; 117° 54′. Cleavage P — ∞. Traces in the direction of P and P — 1 = 114° 51′; 99° 11′. Streak brown. Hardness = 5.0,—5.5. Sp. gr. = 4.7,—4.8.

Description.—Colour brownish-black. Occurs regularly crystallized; also massive, globular, in longish external forms. Lustre glimmering, glistening, and imperfect metallic. Fracture conchoidal and uneven. Opaque.

Constituent Parts.—Red oxide of manganese 98.10, oxygen 0.21, water 0.43, baryta 0.11, silica 0.34. *Turner.*

Geognostic and Geographic Situations.—Occurs in veins in porphyry, in Thuringia, the Hartz, and the United States.

2. *Brachytypous Manganese-Ore, or Braunite.*

Brachytypes Mangan-erz, *Mohs.*

Specific Character.—Pyramidal. P = 109° 53′—108°

39′. Cleavage P, perfect. Streak black, slightly inclining to brown. Hardness = 6.0,—6.5. Sp. gr. = 4.8,—4.9.

Description.—Colour dark brownish-black. Occurs regularly crystallized; also massive. Lustre imperfect metallic and shining. Opaque. Streak black, inclining but slightly to brown. Fracture uneven. Brittle.

Constituent Parts.—Protoxide of manganese 86.94, oxygen 9.85, water 0.95, and baryta 2.26. *Turner.*

Geognostic and Geographic Situations.—Occurs in veins traversing porphyry, at Oehrenstock, near Ilmenau, also at St Marcel, in Piedmont.

Observation.—Named Braunite, in honour of Mr Braun of Gotha.

3. *Uncleavable Manganese-Ore, or Psilomellane.*

Untheilbares Mangan-erz, *Mohs.*—Schwartz-eisenstein, *Werner.*—Fasriger and Dichter Schwartz-braunstein, *Hausmann.*—Dichter Schwarz Mangan-erz, *Leonhard.*—Manganese Oxide-Hydraté Concretionné, *Haüy.*

Specific Character.—Reniform, botryoidal, massive. Uncleavable. Streak brownish-black and shining. Hardness = 5.0,—6.0. Sp. gr. = 4.0,—4.2.

Description.—Colour black, passing into dark steel-grey. Occurs massive, in roundish and longish extended forms. Lustre imperfect metallic, shining or glimmering. Fracture compact and fibrous. Opaque.

Constituent Parts.—Red oxide of manganese 69.79, oxygen 7.36, baryta 16.36, silica 0.26, water 6.22. *Turner.*

Geognostic and Geographic Situations.—This ore of manganese occurs in veins and in nests in primitive, transition, and secondary rocks, very frequently associated with brown iron-ore and grey manganese-ore. In Devonshire and Cornwall, it occurs in botryoidal and stalactitic masses.

It is one of the ores found at Ilefeld, in the Hartz, and is well known in other mining districts in Germany.

4. *Prismatoidal Manganese-Ore, or Grey Manganese-Ore.*

Prismatoidisches Mangan-erz, *Mohs.*—Grauer Branstein, *Werner.*—Manganite, *Haidinger.*—Manganese oxydé, *Haüy.*—Hydrated Deutoxide of Manganese, *Turner.*

Specific Character.—Prismatoidal. Prismatic. $\bar{P}r = 114° 19'$; $P + \infty = 99° 40'$. Cleavage, $Pr + \infty$, very perfect. $P - \infty$, $P + \infty$, less perfect. Streak brown. Hardness = 3.5,—4.0. Sp. gr. = 4.3,—4.4.

Description.—Colour dark steel-grey, inclining to iron-black. Occurs regularly crystallized; in granular, fibrous, and radiated distinct concretions. Lustre shining, glimmering, and metallic. Fracture conchoidal and earthy. Brittle. Opaque.

Constituent Parts.—Red oxide of manganese 86.85, oxygen 3.05, water 10.10. *Turner.*

Geognostic and Geographic Situations.—Occurs in veins and imbedded masses in primitive, transition, and secondary rocks. Its chief locality in Scotland is the vicinity of Aberdeen. Occurs in different parts of England, and also in Ireland. The finest crystallized varieties are found at Ilefeld in the Hartz.

5. *Prismatic Manganese-Ore. Pyrolusite. Common Ore of Manganese.*

Prismatisches Mangan-erz, *Mohs.*—Anhydrous Peroxide of Manganese. *Turner.*

Specific Character.—Prismatic. $P + \infty = 93° 40'$. Cleavage $P + \infty$, $\breve{P}r + \infty$, $Pr + \infty$. Streak black. Soils. Hardness = 2.0,—2.5. Sp. gr. = 4.6,—4.9.

Description.—Colour iron-black, sometimes inclining to bluish-black. Occurs massive, reniform. Lustre metallic. Opaque. Streak black. Soils.

Constituent Parts.—Red oxide of manganese 84.05, oxygen 11.78, water 1.12, baryta 0.53, silica 0.51. *Turner.*

Geognostic and Geographic Situations.—This mineral, the proper ore of manganese, so extensively employed in the arts, occurs in Devonshire and Cornwall, and is well known in the mines of Saxony, France, Hungary, and other countries.[1]

Order IX.—NATIVE METAL.

Metallic. Not black. Hardness = 0.0,—5.0. Sp. gr. = 5.7,—20.0. If grey, is malleable, and sp. gr. = 7.4, and more. If hardness above 4.0, is malleable.

Genus I.—Arsenic.

Rhombohedral. Colour tin-white. Hardness = 3.5. Sp. gr. = 5.7,—5.8.

1. *Rhombohedral Arsenic.*

Gediegen Arsenik, *Werner* and *Mohs.*—Arsenic natif, *Haüy.*

Specific Character.—Rhombohedral. Cleavage R—∞.

Description.—Colour tin-white, which, on exposure, soon tarnishes to black. Occurs massive, in plates, reniform, botryoidal, reticulated, and with impressions. Lustre glistening, glimmering, and metallic. Fracture uneven. Emits, when struck, a ringing sound, and an arsenical odour. Brittle.

Constituent Parts.—Arsenic 97.0, antimony 2.0, iron and water 1.0. *John.*

Geognostic and Geographic Situations.—It occurs in metalliferous veins, particularly where they cross each other, in gneiss, mica-slate, clay-slate, and porphyry; seldomer in

[1] Earthy manganese ore, or black wad of authors, which soils and writes, and has a hardness of about 0.5, is considered to be a hydrate of manganese. Occurs in the manganese mines in Devonshire.

transition and secondary rocks; rarely in beds, and never in large quantity. The mines of Germany, Norway, France, and Russia, afford examples of this mineral.

Genus II.—Tellurium.

Rhombohedral. Colour tin-white. Hardness = 2.0,—2.5. Sp. gr. = 6.1,—6.2.

1. *Rhombohedral Tellurium.*

Gediegen silvan, *Werner*.—Gediegen tellur, *Hausmann*.—Tellur natif ferrifère et aurifère, *Haüy*.

Specific Character.—Rhombohedral. R = 71° 11′. P = 130° 4′; 115° 12′. Cleavage, R — ∞.

Description.—Colour tin-white. Occurs in granular distinct concretions, massive, and disseminated. Lustre shining and metallic. Rather brittle, and easily frangible. Streak unchanged.

Constituent Parts.—Tellurium 92.55, iron 7.20, gold 0.25. *Klaproth.*

Geognostic and Geographic Situations.—Occurs in greywacke, in Transylvania.

Genus III.—Antimony.

Rhombohedral. Prismatic. Colour white, but not inclining to red. Not malleable. Hardness = 3.0,—3.5. Sp. gr. = 6.5,—10.0.

1. *Rhombohedral Antimony. Native Antimony.*

Rhomboedrisches Antimon, *Mohs*.—Gediegen Spiesglas, *Werner*.—Antimoine natif, *Haüy*.

Specific Character.—Rhombohedral. R = 117° 15′.

Cleavage R — ∞, very perfect; traces according to R + 2 and P + ∞. Hardness = 3.0,—3.5. Sp. gr. = 6.5,—6.8

Description.—Colour tin-white. Occurs regularly crystallized; in granular and lamellar concretions; massive, disseminated, and reniform. Lustre splendent and metallic. Rather brittle.

Constituent Parts.—Consists of antimony, with small portions of iron, silver, and arsenic.

Geognostic and Geographic Situations.—Occurs in metalliferous veins in primitive rocks in Sweden, and in the mountains of Hanover, Dauphiny, Hungary, Brazil, and Mexico.

2. *Prismatic Antimony, or Antimonial Silver.*

Prismatisches Antimon, *Mohs.*—Spiesglas Silber, *Werner.*—Argent antimonial, *Haüy.*

Specific Character.—Prismatic. P + ∞ = nearly 120°. Cleavage, P — ∞ . $\breve{P}r$. Less distinct P + ∞. (Fig. 27, 34, 30.) Hardness = 3.5. Sp. gr. = 8.9,—10.0.

Description.—Colour between silver and tin-white. Occurs regularly crystallized, and massive. Lustre splendent and metallic. Sectile.

Constituent Parts.—Silver 84.76, antimony 16.24. *Klaproth.*

Geognostic and Geographic Situations.—Occurs in veins in primitive and transition rocks in Furstenberg, Salzburg, Hartz, and Spain.

Genus IV.—Bismuth.

Tessular. Easily cleavable. Colour silver-white, inclining to red. Hardness = 2.0,—2.5. Sp. gr. = 9.6,—9.8.

1. *Octahedral Bismuth.*

Octaedrisches Wismuth, *Mohs.*— Gediegen Wismuth, *Werner.*—Bismuth natif, *Haüy.*

Specific Character. Tessular. Combination semi-tessular, with inclined planes. Cleavage octahedral, perfect.

Description.—Colour silver-white. Occurs regularly crystallized; massive, disseminated, dentiform, and in leaves with a plumose streaked surface. Lustre splendent and metallic. Sectile, almost malleable.

Constituent Parts.—It is the pure metal, sometimes containing a small quantity of arsenic.

Geognostic and Geographic Situations.—Occurs in veins in primitive rocks, as gneiss, granite, mica-slate, and clay-slate, in Cornwall, Germany, France, Norway, &c.

Genus V.—Mercury.

Tessular. Liquid. Not cleavable. Colour white. Not malleable. Hardness = 0.0,—3.0. Sp. gr. = 10.5,—15.0.

1. *Liquid Native Mercury.*

Flüssiges Mercur, *Mohs.*—Gediegen Quecksilber, *Werner.*—Mercur Natif, *Haüy.*

Specific Character.—Liquid. Colour tin-white. Hardness = 0. Sp. gr. = 12.0,—15.0.

Description.—Colour tin-white. Liquid. Lustre splendent and metallic. Opaque.

Geognostic and Geographic Situations.—Rarely in primitive and transition rocks. More frequently in rocks of the secondary class, as sandstone. Deux-Ponts, Idria, and other European mining districts, afford examples of this metal.

2. *Dodecahedral Mercury, or Native Amalgam.*

Dodecaedrisches Mercur, *Mohs.*—Natürlich Amalgam, *Werner.*—Mercur Argental, *Haüy.*

Specific Character.—Tessular. Silver-white. Hardness = 1.0,—3.5. Sp. gr. = 10.5,—14.0.

Constituent Parts.—Mercury 72.5, silver 27.5. *Cordier.*

Description.—Colour silver-white. Occurs regularly crystallized. Lustre shining and metallic. Fracture uneven. When pressed between the fingers, or cut with a knife, it emits a creaking sound, like artificial amalgam.

Geognostic and Geographic Situations.—Occurs in Deux-Ponts, and other mercury mines, along with cinnabar.

Genus VI.—Silver.

Tessular. Uncleavable. Colour silver-white. Malleable. Hardness = 2.5,—3.0. Sp. gr. 10.0,—11.0.

1. *Hexahedral Silver.*

Hexaedrisches, Silber, *Mohs.*—Argent Natif, *Haüy.*

Specific Character.—Tessular.

Description.—Colour silver-white, and silver-white inclining to brass-yellow. Occurs regularly crystallized, massive, disseminated, dentiform, filiform, reticulated, and in leaves. Lustre metallic, ranging from splendent to glimmering. Fracture hackly. Opaque. Ductile.

The yellow varieties are named *auriferous native silver*, from their containing a portion of gold; the other varieties *common native silver.*

Geognostic and Geographic Situations.—Common native silver occurs in veins, generally occupying their middle or upper parts; and those veins traverse granite, gneiss, mica-slate, clay-slate, hornblende-slate, syenite, and porphyry, in primitive mountains, and greywacke in transition mountains. It rarely occurs in secondary rocks, as in sandstone. The mines of Cornwall, Saxony, Hungary, Mexico, afford this mineral in all its forms. The auriferous native silver was formerly found in the mines of Kongsberg in Norway; at present it is met with in those of Schlangenberg in Siberia.

Genus VII.—Gold.

Tessular. Colour gold-yellow. Hardness 2.5,—3.0. Sp. gr. = 12.0,—20.0.

1. *Hexahedral Gold.*

Hexaedrisches Gediegen Gold, *Mohs.*—Gediegen Gold, *Werner.*—Or natif, *Haüy.*

Specific Character.—Tessular. No cleavage.

Description.—Colours gold-yellow and brass-yellow. Occurs regularly crystallized, massive, disseminated, in flakes, in leaves, reticulated, capillary, dentiform, and in grains. Lustre shining, glistening, and metallic. Fracture hackly. Opaque. Ductile.

The gold-yellow varieties are named *gold-yellow gold;* the brass-yellow varieties, *brass-yellow gold;* those varieties in which the brass-yellow verges on steel-grey, are named *greyish-yellow gold;* and, lastly, the pale brass-yellow, inclining to silver-white, varieties, are named *electrum* or *argentiferous gold.*

Geognostic and Geographic Situations.—Occurs in veins, and disseminated in granite, syenite, gneiss, mica-slate, hornblende-slate, porphyry, greywacke, clay-slate, &c.; also in various alluvial deposites. The mines of Germany, Hungary, and America, afford examples of the various mineralogical and geological relations of this important mineral.

Genus VIII.—Iridium.

Rhombohedral. Easily cleavable. Colour pale steel-grey. Ductile. Hardness = 4.5,—5.0. Sp. gr. = 19.0,—20.0.

1. *Rhombohedral Iridium. Alloy of Iridium and Osmium.*

Rhomboedrisches Iridium, *Mohs.*—Osmium-Iridium, *Leonhard.*—Iridium Osmie, *Haüy.*

Specific Character.—Rhombohedral. Cleavage R — ∞, perfect.

Description.—Colour pale steel-grey. Occurs in grains, rarely in six-scaled prisms. Lustre metallic. Cleavage easily obtained. Opaque.

Constituent Parts.—Osmium 24.5, iridium 72.9, iron 2.6. *Thomson.*

Geognostic and Geographic Situations.—It occurs along with platina in the province of Choco in South America, and also in the Uralian Mountains.[1]

Genus IX.—Palladium.

Tessular. Colour steel-grey. Ductile. Hardness = 4.5, —5.0. Sp. gr. = 11.5,—12.5.

1. *Octahedral Palladium.*

Octaedrisches Palladium, *Mohs.*

Specific Character.—Tessular. Uncleavable.

Description.—Colour steel-grey. Occurs in grains, having a fibrous structure. Lustre metallic. Ductile, and very malleable.

Geognostic and Geographic Situations.—Occurs along with platina and iridium in Minas Geras, in Brazil.

Genus X.—Platina.

Tessular. Uncleavable. Colour steel-grey. Ductile. Hardness = 4.0,—4.5. Sp. gr. = 16.0,—20.0.

[1] Breithaupt notices a true native iridium, having a specific gravity of 23.55, from Siberia.

1. *Hexahedral Platina.*

Gediegen Platin, *Werner.*—Hexahedrisches Platin, *Mohs.* —Platin natif, *Haüy.*

Specific Character.—Tessular. In grains.

Description.—Colour steel-grey. Occurs in small flat or angular grains, seldomer in blunt-edged pieces. Lustre metallic, and shining. Fracture hackly. No cleavage, Ductile.

Constituent Parts.—Grains from Columbia. Platina 84.30, rhodium 3.46, palladium 1.06, iridium 1.46, osmium 1.03, copper 0.74, iron 5.31. *Berzelius.*

Geognostic and Geographic Situations.—It occurs in blunt-edged pieces and grains along with iridium, palladium, gold, magnetic iron-sand, &c. in the alluvial deposites of Brazil, Choco, and Barbaco in South America. It also occurs, and rather abundantly, at Joetsk in the Government of Perm in Siberia.

Genus XI.—Iron.

Tessular. Colour pale steel-grey. Hardness 4.5. Sp. gr. = 7.4,—7.8.

1. *Octahedral Iron.*

Octaedrisches Eisen, *Mohs.*—Gediegen Eisen, *Werner.* —Fer natif, *Haüy.*

Specific Character.—Tessular. No cleavage.

Description.—Colour pale steel-grey. Occurs ramose, and disseminated in meteoric stones. Lustre glimmering, glistening, and metallic. Fracture hackly.

Constituent Parts.—1. *Atacama.* Iron 93.40, nickel 6.62, cobalt 0.53. *Turner.*—2. *Siberia.* Iron 98.5, nickel 1.5. *Klaproth.*—3. *Agram.* Iron 96.5, nickel 3.5. *Klaproth.*

Geognostic and Geographic Situations.—It is a meteoric production, and has been observed to fall from fireballs in Europe, Asia, and America.

Genus XII.—Copper.

Tessular. Colour copper-red. Hardness = 2.5,—3.0. Sp. gr. = 8.4,—8.9.

1. *Octahedral Copper.*

Octaedrisches Kupfer, *Mohs.*—Gediegen Kupfer, *Werner.*—Cuivre natif, *Haüy.*

Specific Character.—Tessular. No cleavage.

Description.—Colour copper-red. Occurs regularly crystallized; massive, dendritic, capillary, botryoidal, and ramose. Lustre glistening and metallic. Fracture hackly. Streak shining but unchanged in colour. Ductile.

Geognostic and Geographic Situations.—Occurs in granite, gneiss, mica-slate, clay-slate, primitive limestone, syenite, serpentine, greywacke, secondary limestone, sandstone, and generally in small veins; also in grains, and sometimes in blocks many pounds weight, in alluvial districts. It occurs in serpentine in Shetland, and in the copper mines of Cornwall. Large masses are met with in alluvial districts in the northern parts of North America.

Order X.—PYRITES.

Metallic. Streak black. Hardness = 3.0,—6.5. Sp. gr. = 4.1,—7.7. If hardness = 4.5 and less, the sp. gr. is less than 5.3. If sp. gr. = 5.3 and less, the colour is yellow or red.

Genus I.—Nickel Pyrites, or Copper-Nickel. Arsenical Nickel.

Prismatic. Colour copper-red. Hardness = 5.0,—5.5. Sp. gr. = 7.5,—7.7.

1. *Prismatic Nickel Pyrites.*

Prismatischer Nickel-kies, *Mohs.*—Kupfer-Nickel, *Werner.*—Nickel Arsenical, *Haüy.*

Specific Character.—Prismatic. Cleavage very imperfect.

Description.—Colour copper-red. Rarely crystallized; in granular distinct concretions; massive, disseminated, reticulated, dendritic, fruticose, globular, and botryoidal. Lustre shining and metallic. Fracture conchoidal. Brittle.

Constituent Parts.—Nickel 44.21, arsenic 54.73.—*Stromeyer.*

Geognostic and Geographic Situations.—Occurs in silver and cobalt veins in gneiss, mica-slate, clay-slate, and syenite; also in veins in secondary rocks, particularly bituminous marl-slate. In Scotland it is met with at Leadhills and Wanlockhead, and in the coal-field of West Lothian.

* *Nickel Ochre.*—Colour apple-green. Occurs as a thin coating, seldom massive, and disseminated. Dull. Fracture uneven or earthy. Opaque, or translucent on the edges. Soft. Feels meagre. It occurs in mineral veins along with copper-nickel at Alva in Stirlingshire, in Linlithgowshire, and at Leadhills.

** *Black Nickel.*—Colour black. Occurs massive, disseminated, in crusts. Dull. Fracture earthy. Opaque. Soft. Shining in streak. Soils slightly. It occurs in veins in bituminous marl-slate at Reichelsdorf.

Genus II.—Arsenic Pyrites.

Prismatic. Colour not inclining to red. Hardness = 5.0,

—6.0. Sp. gr. = 5.7,—7.4. If colour white or grey, the sp. gr. is under 6.3, or above 7.0.

1. *Axotomous Arsenic Pyrites.*

Axotomer Arsenik-kies, *Mohs.*—Arsenik-kies, *Werner.* —Fer Arsenical, *Haüy.*

Specific Character.—Prismatic. $\bar{P}r = 51° 20'$. $P + \infty = 122° 26'$. Cleavage $P - \infty$, perfect. Less $\breve{P}r = 86° 10'$. Traces in direction of $P + \infty$. (Fig. 27, 30). Hardness = 5.0,—5.5. Sp. gr. = 7.1,—7.4.

Description.—Colour pale steel-grey. Occurs massive. Lustre metallic and shining. Streak greyish-black. Fracture uneven. Brittle.

Constituent Parts.—It is a compound of arsenic and iron.

Geognostic and Geographic Situations.—Occurs with copper-nickel in Styria, with serpentine in Silesia, and also in Hungary.

2. *Prismatic Arsenic Pyrites.*

Prismatischer Arsenik-kies, *Mohs.*—Arsenikkies, *Werner.*—Fer Arsenical, *Haüy.*

Specific Character.—Prismatic. $\bar{P}r - 1 = 145° 26'$; $P + \infty$, $111° 53'$. Cleavage $P - \infty$. $(\breve{P}r + \infty)^3$ $111° 19'$. (Fig. 27, 32.) Hardness = 5.5,—6.0. Sp. gr. = 5.7,—6.2.

Description.—Colour silver-white, inclining to steel-grey. Occurs regularly crystallized; in prismatic concretions, massive and disseminated. Lustre ranges from splendent to glistening, and is metallic. Fracture uneven. Opaque. Brittle. Streak greyish-black.

Constituent Parts.—Iron 36,04, arsenic 42.88, sulphur 21.08. *Stromeyer.*

Geognostic and Geographic Situations.—This mineral occurs in a variety of metalliferous formations in primitive mountains, and also in those of the transition and secondary classes. In Scotland it occurs in secondary rocks in Stirlingshire, and in England in considerable quantity in the mines of Devon and Cornwall.

Genus III.—Cobalt Pyrites.

Tessular. Colour white, inclining to steel-grey or red. Hardness = 5.0,—5.5. Sp. gr. = 6.0,—6.6.

1. *Hexahedral Cobalt Pyrites, or Silver White Cobalt.*

Hexaedrisches Kobalt-kies, *Mohs.*—Glanz Kobold, *Werner.*—Cobalt gris, *Haüy.*

Specific Character.—Tessular. Combination semi-tessular, with parallel faces. Cleavage hexahedral and perfect. Colour white, inclining to red. Hardness = 5.5. Sp. gr. = 6.1,—6.35.

Description.—Colour silver-white, inclining to copper-red. Occurs regularly crystallized; in granular distinct concretions, massive, disseminated, and reticulated. Lustre shining, glistening, and metallic. Fracture uneven or conchoidal. Brittle. Easily frangible. Streak greyish-black.

Constituent Parts.—From *Modum.* Cobalt 33.10, arsenic 43.46, iron 3.23, sulphur 21.08. *Stromeyer*

Geognostic and Geographic Situations.—Occurs in beds, in mica-slate, associated with iron and copper-pyrites, magnetic iron-ore, anthophylite, tourmaline, felspar, &c. in the parish of Modum, in Norway, also at Tunaberg, in Sweden. In some transition districts it is found in veins along with various ores of iron and copper.

2. *Octahedral Cobalt-Pyrites, or Tin-White Cobalt.*

Octaedrischer Cobalt-kies, *Mohs.*—Weisser Speis-kobald, *Werner.*—Cobalt Arsenical, *Haüy.*

Specific Character.—Tessular. Cleavage hexahedral, octahedral, dodecahedral, but indistinct. Tin-white, inclining to steel-grey. Hardness = 5.5. Sp. gr. = 6.4,—6.6.

Description.—Colour tin-white, inclining to steel-grey. Occurs regularly crystallized, and the faces of the crystals frequently cracked; in lamellar and granular concretions; massive, disseminated, reticulated, fruticose, specular. Lustre shining, splendent, and metallic. Fracture uneven, Brittle and easily frangible. Streak greyish-black.

Constituent Parts.—*Riechelsdorf.* Cobalt 20.31, arsenic 74.21, iron 3.42, copper 0.15, sulphur 0.88. *Stromeyer.*

Geognostic and Geographic Situations.—Occurs in veins in primitive and transition rocks; also in old red sandstone, and in copper-slate. Cornwall, Hessia, Thuringia, Hartz, &c. afford many localities of this mineral.

3. *Isometric Cobalt-Pyrites. Sulphuret of Cobalt.*

Isometrisher Cobalt-Kies, *Mohs.*—Cobalt-Kies, *Hausmann.*

Specific Character.—Tessular. Cleavage hexahedral, imperfect. Colour white, inclining to red. Hardness = 5.5. Sp. gr. = 6.3,—6.4.

Description.—Colour white, inclining to red. Occurs massive, in granular concretions, and in cubo-octahedral crystals. Lustre metallic. Fracture uneven or imperfect conchoidal.

Constituent Parts.—*Musen, in Nausau Siegen.* Cobalt 53.35, sulphur 42.25, iron 2.30, copper 0.97. *Wernekink.*—2. *Ridderhyttan, Sweden.* Cobalt 43.20, copper 14.40, iron 3.53, sulphur 38.50. *Hisinger.*

Geognostic and Geographic Situations.—Occurs in gneiss, associated with hornblende and copper-pyrites at

Bastnais, near Ridderhyttan, in Sweden, and at Musen, near Siegen, with heavy spar and sparry iron.

4. *Eutomous Cobalt-Pyrites. Antimonial Nickel.*[1]

Eutomer Kobalt-kies, *Mohs.*—Nickelspiesglasserz, *Hausmann.*—Nickel Arsenical Antimonifere, *Haüy.*

Specific Character.—Tessular. Cleavage hexahedral, perfect. Colour pale steel-grey. Hardness = 5.0,—5.5. Sp. gr. = 6.4,—6.5.

Description.—Colour pale steel-grey, sometimes on exposure becoming blackish. Colour of the streak darker. Occurs massive, disseminated, and in cubo-octahedral crystals, and in granular concretions. Lustre metallic. Fracture uneven. Brittle.

Constituent Parts.—Nickel 27.36, antimony 55.76, sulphur 15.98. *H. Rose.*

Geognostic and Geographic Situations.—Occurs in veins of ironstone and lead-glance, in transition rocks along with sparry iron, cobalt-pyrites, grey copper, and copper-pyrites, in Nassau Siegen in the Westerwald.

Genus IV.—Iron-Pyrites.

Tessular, rhombohedral, prismatic, Yellow, partly inclining to copper-red. Hardness = 3.5,—6.5. Sp. gr. = 4.4,—5.05.

1. *Hexahedral Iron-Pyrites.*

Hexaedrischer Eisen-Kies, *Mohs.*—Gemeiner Schwefel-Kies, *Werner.*—Fer Sulphuré, *Haüy.*

Specific Character.—Tessular. Combination semi-tessular with parallel faces. Cleavage hexahedral and octahedral.

[1] This species of cobalt pyrites, although containing no cobalt, is placed here on account of its natural-historical relations to the other species of the genus.

Colour bronze-yellow. Hardness = 6.0,—6.5. Sp. gr. = 49,—5.05..

Description.—Colour bronze-yellow, sometimes inclining to steel-grey. Occurs regularly crystallized in various tessular forms; in granular concretions; massive, disseminated, globular, and cellular. Lustre ranges from shining to glimmering, and is metallic. Fracture uneven and conchoidal. Opaque. Brittle.

Constituent Parts.—Iron 47.85, sulphur 52.15. *Hatchett.*

Geognostic and Geographic Situations.—The *cellular varieties* are rare, and hitherto have been met with principally in Saxony; while the others occur in all countries, and more or less extensively distributed through rocks of every description, from those of the oldest primitive, to the newest alluvial formation.

2. *Prismatic Iron-Pyrites or Radiated Iron-Pyrites.*

Prismatischer Eisenkies, *Mohs.*—Fer Sulphuré Blanc, *Haüy.*—Wasser-Kies. Strahl-Kies. Kam-kies. Leber-Kies. Spär-Kies. Zell-Kies. *Werner.*

Specific Character.—Prismatic. $\breve{P}r = 114° \ 19'$. $\bar{P}r = 106° \ 36'$. $P + \infty = 98° \ 13'$. Cleavage $\bar{P}r$. Traces according to $P + \infty$. (Fig. 30.) Colour bronze-yellow. Hardness = 6.0,—6.5. Sp. gr. = 4.65,—4.9.

Description.—Colour bronze-yellow, inclining sometimes to steel-grey, or to brass-yellow. Occurs regularly crystallized; in radiated, granular, and lamellar concretions; massive, dendritic, reniform, globular, stalactitic, botryoidal, fruticose, and with impressions. Lustre varies from glistening to glimmering, and is metallic. Opaque. Brittle. Easily frangible.

The varieties in radiated concretions are named *radiat-*

ed-pyrites; those in which the colour inclines to brass-yellow, and which, on exposure to air, acquire a brown tarnish, are named *hepatic-pyrites;* those in spear-shaped twin and triple crystals, *spear-pyrites;* and lastly, those in which the crystals are so aggregated as to have the form of the comb of the cock, are named *cockscomb-pyrites.*

Constituent Parts.—Iron 45.66, sulphur 54.34. *Hatchett.*

Geognostic and Geographic Situations.—This species of iron-pyrites occurs more frequently and abundantly in newer than in older formations. The newest secondary formations, and those of the alluvial class, both in this island and on the continent of Europe, afford numerous localities of the radiated varieties. The spear-pyrites is met with in Bohemia and Saxony; and the cockscomb-pyrites in veins in Derbyshire, and in some mines in Saxony.

3. *Rhombohedral Iron-Pyrites, or Magnetic-Pyrites.*

Rhomboedrischer Eisenkies, *Mohs.*—Magnetkies, *Werner.*—Fer Sulphuré Ferrifère, *Haüy.*

Specific Character.—Rhombohedral. Rhombohedron unknown. Combination di-rhombohedral. Cleavage, $P - \infty$, perfect. Less distinct $P + \infty$. Colour bronze-yellow, inclining to copper-red. Hardness = 3.5,—4.5. Sp. gr. 4.4,—4.7.

Description.—Colours intermediate between bronze-yellow and copper-red. Occurs rarely crystallized; in granular concretions, also massive and disseminated. Lustre ranges from splendent to glistening, and is metallic. Fracture conchoidal and uneven. Opaque. Brittle. Easily frangible.

Constituent Parts.—Iron 63.5, sulphur 36.5. *Hatchett.*

Geognostic and Geographic Situations.—This mineral occurs disseminated in primitive and transition rocks, and also disposed in beds in rocks of the same classes, in Scotland, England, Saxony, &c.

Genus V.—Copper-Pyrites.

Tessular. Pyramidal. Colours brass-yellow and copper-red. Hardness = 3.5,—4.0. Sp. gr. = 4.1,—5.1.

1. *Pyramidal Copper-Pyrites, or Yellow Copper-Pyrites.*

Pyramidaler Kupfer-kies, *Mohs.*—Kupfer-kies, *Werner.*—Cuivre pyriteux, *Haüy.*

Specific Character.—Pyramidal. Pyramid 109° 53′; 108° 40′. Combination hemi-pyramidal, with inclined faces. Cleavage, P + 1 = 101° 49′; 126° 11′. Colour brass-yellow. Hardness = 3.5,—4.0. Sp. gr. = 4.1,—4.3. Colour brass-yellow. Occurs massive, disseminated, globular, reniform, botryoidal, stalactitic; in granular concretions. Lustre shining and metallic. Fracture uneven or conchoidal. Streak greenish-black and feebly shining.

Constituent Parts.—Copper 33.12, iron 30.00, sulphur 36.52, silica 0.39, *Rose.*

Geognostic and Geographic Situations.—This species of pyrites is found in primitive, transition, and secondary rocks, not only in veins, but also in beds, and in vast imbedded masses. The copper-mines in England afford it in great variety and abundance; it occurs also in Scotland, but in smaller quantities.

2. *Octahedral Copper-Pyrites. Variegated or Purple Copper.*

Bunt Kupfererz, *Werner.*—Octaedrischer Kupfer-Kies, *Mohs.*—Cuivre-Pyriteux-hepatique, *Haüy.*

Specific Character.—Tessular. Cleavage octahedral, very imperfect. Colour copper-red. Hardness = 3.0. Sp. gr. = 4.9,—5.1.

Description.—Colour between copper-red and pinchbeck-brown, but soon acquires a variegated tarnish. Occurs in cubes truncated on the angles; also massive. Lustre shining and metallic. Fracture conchoidal.

Constituent Parts.—From Killarney. Copper 61.07, sulphur 23.75, iron 14.00, silica 0.50. *Phillips.*

Geognostic and Geographic Situations.—Occurs in primitive, transition, and secondary rocks. The finest British specimens are from the coppermines of Cornwall. Ross Island in Killarney in Ireland, also affords this ore, as do the copper mines in Norway, Hessia, Silesia, the Bannat, and Siberia.

Order XI.—GLANCE.

Metallic. Colour and streak grey, black. Hardness = 1.0,—4.0. Sp. gr. = 4.2,—7.6. If cleavage monotomous, and sp. gr. under 5.0, the colour is lead-grey. If hardness = 2.5 and less, the sp. gr. is above 5.0, or the colour is lead-grey. If sp. gr. above 7.4, the colour is lead-grey.

Genus I.—Copper-Glance.

Tessular, prismatic. Colour blackish lead-grey; steel-grey, sometimes inclining to yellow; black. Cleavage imperfect, not axotomous. Hardness = 2.5,—4.0. Sp. gr. = 4.3,—5.8.

1. *Tetrahedral Copper-Glance. Grey Copper. Black Copper.*

Tetradrischer Kupfer-Glanz, *Mohs.*—Fahlerz. Schwarzerz, *Werner.*—Cuivre gris, *Haüy.*

Specific Character.—Tessular. Combination semi-tessular with inclined faces. Cleavage octahedral. Colour steel-grey,...iron-black. Hardness = 3.0,—4.0. Sp. gr. = 4.5,—5.2.

Description.—Colours steel-grey and iron-black. Occurs

regularly crystallized, massive, and disseminated. Lustre externally splendent and metallic, internally shining, or glistening, and metallic. Fracture conchoidal or uneven. Opaque. Brittle. Easily frangible.

The grey varieties are named *grey copper;* the black, *black copper.*

Constituent Parts.—Fahlerz. Copper 48.0, arsenic 14.0, sulphur 10.0, iron 25.5, silver 0.5. *Klaproth. Schwartzerz.* Copper 40.25, arsenic 0.75, antimony 23.00, sulphur 18.50, iron 13.50, silver 0.30. *Klaproth.*

Geognostic and Geographic Situations.—The grey varieties occur in veins in transition granite, and syenite, at Fassney Burn, in East Lothian; at Airthrie, in Stirlingshire; in Ayrshire, and in Devonshire. The black varieties are found in transition rocks, at Clausthal in the Hartz.

2. *Prismatoidal Copper-Glance.*

Prismatoidischer Kupfer-glanz, *Mohs.*

Specific Character.—Prismatic. Pyramid unknown. Cleavage, $\breve{P}r + \infty$. (Fig. 29.) Colour blackish lead-grey. Brittle. Hardness = 3.0. Sp. gr. = 5.7,—5.8.

Description.—Colour blackish lead-grey. Crystals in oblique four-sided prisms. Lustre shining and metallic.

Geognostic and Geographic Situations.—Occurs along with sparry iron at St Gertraud, near Wolfsberg, in the valley of Lavan, in Carinthia.

3. *Prismatic Copper-Glance, or Vitreous Copper.*

Prismatischer Kupfer-Glanz, *Mohs.*—Kupferglas, *Werner.*—Cuivre Sulphuré, *Haüy.*

Specific Character.—Prismatic. $\breve{P}r = 119° 35'$. $(\breve{P} + \infty.)^2 = 63° 48'$. Cleavage $\breve{P}r$. Colour blackish lead-grey. Very sectile. Hardness = 2.5,—3.0. Sp. gr. = 5.5,—5.8.

Description.—Colour blackish lead-grey. Occurs regularly crystallized, also in granular concretions, and massive. Lustre glistening, glimmering and metallic. Fracture conchoidal and uneven. Opaque. Sectile.

Constituent Parts.—Copper 84.0, sulphur 12.0, iron 4.0. *Chenevix.*

Geognostic and Geographic Situations.—Occurs in veins at Fassney Burn; in the rocks of Fair Isle; in Yorkshire, Caernarvonshire, and Cornwall.

4. *Diprismatic Copper-Glance, or Bournonite.*

Schwarz-Spiesglaserz, *Werner.*—Diprismatischer Kupfer-Glanz, *Mohs.*—Plomb Sulphure Antimonifère, and Antimoin Sulphuré Plombo-Cuprifère, *Haüy.*—Endellione.

Specific Character.—Prismatic. $\breve{P}r = 93^\circ\ 40'$; $\bar{P}r - 1 = 87^\circ\ 8'$; $(\breve{P} + \infty)^2 = 96^\circ\ 31'$. Cleavage $\breve{P}r + \infty$; $\bar{P}r + \infty$. First the more perfect of the two. Colour steel-grey, inclining to lead-grey or iron-black. Brittle. Hardness = 2.5,—3.0. Sp. gr. = 5.7,—5.8.

Description.—Colour steel-grey, inclining to lead-grey, or iron-black. Occurs massive and regularly crystallized. Lustre splendent and metallic. Fracture conchoidal or uneven. Opaque. Streak unchanged. Brittle and very easily frangible.

Constituent Parts.—From Endellion. Lead 42.62, antimony 24.23, copper 12.80, iron 1.20, sulphur 17.00. *Hatchett.*

Geognostic and Geographic Situations.—The finest specimens of this mineral are found in the mines of Neudorf, in the Hartz; it was, we believe, first found in the parish of Endellion, near Redruth, in Cornwall, and was described by Bournon under the name Endellionite, but we, in compliment to Bournon, named it Bournonite.

5. *Hexahedral Copper-Glance.* *Tin-Pyrites.*

Zinnkies, *Werner*.—Hexadrischer Kupfer-Glanze, *Mohs*. —Sulphuret of Copper and Tin.

Specific Character.—Tessular. Cleavage hexahedral, dodecahedral. Colour steel-grey, inclining to brass-yellow. Brittle. Hardness = 4.0. Sp. gr. = 4.3,—4.4.

Description.—Colour steel-grey, inclining more or less to brass-yellow. Streak black. Occurs massive and crystallized in cubes. Lustre metallic. Fracture uneven. Brittle. Opaque.

Constituent Parts.—Tin 34.0, copper 36.0, iron 2.0, sulphur 25.0. *Klaproth*.

Geognostic and Geographic Situations.—It has hitherto been found only in metalliferous veins in Cornwall, principally at Huel Rock, in the parish of St Agnes, where it is associated with blende, copper-pyrites, and other minerals.

* Tennantite, *Phillips*.

Description.—Colour blackish lead-grey. Streak dark-reddish grey. Tessular. Cleavage dodecahedral, but imperfect. Lustre metallic. Fracture uneven. Brittle. Hardness = 4.0. Sp. gr. = 4.3,—4.4.

Constituent Parts.—Copper 45.32, arsenic 11.84, iron 9.26, sulphur 28.74, silica 5.00. *Phillips*.

Geognostic and Geographic Situations.—Hitherto this mineral has been found only in Cornwall, where it occurs in copper veins, that traverse granite and clay-slate, in the mines near Redruth and St Day.

Genus II.—Silver-Glance, or Vitreous Silver.

Silber-Glanz, *Mohs.*

Tessular. Colour blackish lead-grey. Hardness = 2.0, —2.5. Sp. gr. = 6.9,—7.2.

1. *Hexahedral Silver-Glance.*

Hexaedrischer Silber-glanz, *Mohs.*—Glaserz, *Werner.*—Argent Sulphuré, *Haüy.*

Specific Character.—Tessular. Cleavage dodecahedral, but very impesfect. Malleable.

Description.—Colour blackish lead-grey. Occurs regularly crystallized, most frequently in cubo-octahedrons; massive, disseminated, in plates, dentiform, filiform, capillary, reticulated, dendritic, stalactitic, and with impressions. Lustre shining, glistening, and metallic. Fracture uneven or conchoidal. Completely malleable. Flexible, but not elastic.

Constituent Parts.—Silver 85.0, sulphur 15.0. *Klaproth.*

Geognostic and Geographic Situations.—It is one of the most common of the ores of silver. It was formerly met with at Airthrie in Stirlingshire, and is still found in the mines of Cornwall.

Genus III.—Galena, or Lead-Glance.

Blei-Glanz, *Werner.*

Tessular. Colour pure lead-grey. Hardness = 2.5. Sp. gr. = 7.4,—7.6.

1. *Hexahedral Galena, or Lead-Glance.*

Hexaedrischer Bleiglanz, *Mohs.*—Bleiglanz, *Werner.*—Plomb Sulphuré, *Haüy.*

Specific Character.—Tessular. Cleavage hexahedral, perfect.

Description.—Colour lead-grey. Occurs regularly crystallized, frequently in cubes and cubo-octahedrons; in granular, prismatic, and lamellar concretions; massive, disseminated, specular, reticulated, botryoidal, and corroded. Lustre splendent to glimmering and metallic. Fracture even, or flat conchoidal. Fragments cubical. Sectile. Uncommonly easily frangible.

The variety with glimmering lustre, and even or conchoidal fracture, is named *compact galena.*

Constituent Parts.—Lead 85.13, sulphur 13.02. *Thomson.* A good series of analyses of this ore much wanted.

Geognostic and Geographic Situations.—This mineral, the species from which all the lead of commerce is obtained, occurs in every lead mine, whether in primitive, transition, or secondary rocks.

Genus IV.—Tellurium-Glance, or Black Tellurium.

Tellur-Glanz, *Mohs.*

Pyramidal. Colour blackish lead-grey. Cleavage monotomous, perfect. Hardness = 1.0,—1.5. Sp. gr. = 7.0,—7.2.

1. *Pyramidal Tellurium-Glance.*

Pyramidaler Tellur-Glanz, *Mohs.*—Nagyagerz, *Werner.* —Tellure Natif Auro-Plombifère, *Haüy.*

Specific Character.—Pyramidal. P = 96° 43′; 140° 0′. Cleavage P — ∞.

Description.—Colour blackish lead-grey. Occurs regularly crystallized; massive, disseminated, and in leaves. Lustre splendent and metallic. Sectile. Folia flexible.

Constituent Parts.—Tellurium 32.2, lead 54.0, gold 9.0, silver 0.5, copper 1.3, sulphur 3.0. *Klaproth.*

Geognostic and Geographic Situations.—Occurs in veins that traverse porphyry, in Transylvania.

Genus V.—Molybdena, or Molybdena-Glance.

Molybdän-Glanz, *Mohs.*

Rhombohedral. Colour pure lead-grey. In thin leaves, easily flexible. Hardness = 1.0,—1.5. Sp. gr. = 4.4,—4.6.

1. *Rhombohedral Molybdena.*

Rhomboedrischer Molybdän, *Mohs.*—Wasserblei, *Werner.*—Molybdene Sulphuré, *Haüy.*

Specific Character.—Rhombohedral. Combination di-rhombohedral. Cleavage R — ∞, very perfect.

Description.—Colour fresh lead-grey. Occurs crystallized in very flat six-sided prisms; massive, disseminated, and in plates; in granular concretions. Lustre splendent or shining and metallic. Sectile, approaching to malleable. On porcelain it leaves a greenish streak, on paper a lead-grey streak.

Constituent Parts.—Molybdena 60.0, sulphur 40.0. *Bucholz.*

Geognostic and Geographic Situations.—Occurs imbedded in granite and syenite at Peterhead; in chlorite-slate in Glenelg; in granite and syenite in Corybuy, at the head of Loch Creran, and in various mines in Cornwall.

* *Molybdena Ochre.*

The sulphur-yellow mineral which sometimes incrusts molybdena is named *molybdena ochre.*

Genus VI.—Bismuth-Glance.

Prismatic. Colour pure lead-grey. Hardness = 2.0,—2.5. Sp. gr. = 6.1,—6.4.

1. *Prismatic Bismuth-Glance.*

Prismatischer Wismuth-Glanz, *Mohs.*—Wismuth-Glanz, *Werner.*—Bismuth sulphuré, *Haüy.*

Specific Character.—Prismatic. Pyramid unknown. Cleavage $P - \infty$; $P + \infty$, imperfect. $P\breve{r} + \infty$. $P\bar{r} + \infty$, larger, more perfect. (Fig. 30, 29, 28.)

Description.—Colour pale lead-grey. Occurs regularly crystallized; in granular and radiated concretions; massive and disseminated. Internally splendent and metallic. Fracture foliated or fibrous. Soils. Streak unchanged. Brittle, inclining to sectile. Easily frangible.

Constituent Parts.—Bismuth 80.98, sulphur 18.72. *Rose.*

Geognostic and Geographic Situations.—Occurs in veins in Cornwall; associated with molybdena and apatite at Caldbeck Fell, in Cumberland.

* *Bismuth-Ochre.*

Wismuth-ocker, *Werner.*

The yellow, grey, or green mineral which sometimes accompanies bismuth-glance, is the bismuth ochre of mineralogists.

Genus VII.—Antimony-Glance.

Antimon-Glanz, *Mohs.*

Prismatic. Colour lead-grey, not blackish; steel-grey. Cleavage perfect. Hardness = 1.5,—2.5. Sp. gr. = 4.2, —5.8. If the specific gravity is less than 5.3, the hardness = 2.0; easily broken when in thin plates. If sp. gr. above 5.3, the colour is steel-grey.

1. *Prismatic Antimony-Glance, or Graphic Tellurium.*

Prismatischer Antimon-glanz, *Mohs.*—Schrift-erz, *Werner.*—Tellure natif auro-argentifère, *Haüy.*

Specific Character.—Prismatic. $P + \infty = 94° \ 20'$. Cleavage $\breve{P}r + \infty$, perfect. Less distinct $\bar{P}r + \infty$. (Fig. 29, 28.) Colour pure steel-grey. Hardness = 1.5,—2.0. Sp. gr. = 5.7,—5.8.

Description.—Colour steel-grey. Occurs regularly crystallized; massive, disseminated, and in leaves. Externally splendent, internally glistening, and metallic. Fracture uneven. Sectile.

Constituent Parts.—Tellurium 61.35, gold 28.36, silver 10.29. *Klaproth.*

Geognostic and Geographic Situations.—Occurs in porphyry veins in Transylvania.

2. *Prismatoidal Antimony-Glance, or Grey Antimony.*

Grau Spiesglaserz, *Werner.*—Prismatoidischer Antimon-Glanz, *Mohs.*—Antimoine sulphuré, *Haüy.*

Specific Character.—Prismatic. $P = 109° \ 16'$; $108° \ 10'$; $110° \ 59'$. $P + \infty = 90° \ 45'$. Cleavage $\breve{P}r + \infty$, perfect. Less distinct, $P - \infty$. $P + \infty$. $\bar{P}r + \infty$. (Fig. 29, 27, 30, 28.) Colour lead-grey. Hardness = 2.0. Sp. gr. = 4.2, —4.7.

Description.—Colour lead-grey. Occurs regularly crystallized; in radiated, fibrous, and granular distinct concretions. Lustre ranges from glistening to splendent, and is metallic. Fracture even and uneven. Rather brittle, and easily frangible.

Constituent Parts.—Antimony 74.06, sulphur 25.94. *Davy.*

Geognostic and Geographic Situations.—Occurs in veins

that traverse greywacke, at Westerhall, in Dumfriesshire, and in Banffshire in primitive rocks. Being the ore from which the antimony of commerce is obtained, it occurs in all antimony mines.

3. *Axotomous Antimony-Glance, or Jamesonite.*

Axotomer Antimon-Glanz, *Mohs.*—Jamesonite, *Haidinger.*—Triple Sulphuré d'Antimoine, Plomb et Cuivre; Endellione, *Bournon.*—Antimoine Sulphuré, *Haüy.*

Specific Character.—Prismatic. $P + \infty = 101° 20'$. Cleavage $P - \infty$, very perfect. $P + \infty$, $\breve{P}r + \infty$, imperfect. (Fig. 27.) Colour steel-grey. Hardness = 2.0,—2.5. Sp. gr. = 5.5,—5.8.

Description.—Colour steel-grey. Occurs massive, and in parallel or scopiform prismatic concretions. Lustre externally shining, internally glistening and metallic. Fracture uneven or conchoidal. Opaque. Sectile.

Constituent Parts.—Antimony 34.40, lead 40.75, sulphur 22.15, iron 2.30, with traces of zinc and copper. *Rose.*

Geognostic and Geographic Situations. It was first noticed in veins traversing clay-slate in Cornwall, and since in Hungary and Siberia.

4. *Peritomous Antimony-Glance.*

Peritomer Antimon-Glanz, *Mohs.*—Schilfglaserz, *Freiesleben.*—Argent Sulphuré Antimonifère et Cuprifère, *Levy.*—Sulphuret of Antimony and Silver. *Phillips.*

Specific Character.—Prismatic. $\breve{P}r = 130° 8'$. $P + \infty = 100° 0'$. Cleavage $P + \infty$, perfect. Colour pale steel-grey. Hardness = 2.0,—2.5. Sp. gr. = 5.5,—5.6.

Description.—Colour pale steel-grey, inclining to silver-white. Occurs massive. Surface striated longitudinally. Lustre metallic.

Constituent Parts.—Conjectured to be a compound of antimony, sulphur, silver, and copper.

Geognostic and Geographic Situations.—Occurs in the Himmelsfurst mine, near Freyberg, in Saxony, also at Kapnick, in Transylvania.

Genus VIII.—Melane-Glance.

Prismatic. Colour iron-black. Hardness = 2.0,—2.5. Sp. gr. = 5.9,—6.4.

1. *Prismatic Melane-Glance. Brittle Silver-Glance. Black Silver-Glance.*

Prismatischer Melan-Glance, *Mohs.*—Sprödglaserz, *Werner.*—Argent Antimoine-Sulphuré Noire, *Haüy.*

Specific Character.—Prismatic. $P = 130° 16'; 104° 19'; 96° 7'$. $\breve{P}r = 115° 39$. $(\breve{P} + \infty)^2 = 72° 13'$ Cleavage $(\breve{P} + \infty)^2$, $\breve{P}r + \infty$. Imperfect. (Fig. 30, 29.)

Description.—Colour between iron-black and dark lead-grey. Occurs regularly crystallized and disseminated. Lustre externally splendent, internally shining, and metallic. Fracture conchoidal, and uneven. Streak unchanged. Opaque. Sectile.

Constituent Parts.—Silver 68.54, antimony 14.68, copper and arsenic 0.64, sulphur 16.42. *Rose.*

Geognostic and Geographic Situations.—Occurs in veins in primitive rocks in Hungary, Saxony, Bohemia, Peru, and Mexico.

Order XII.—BLENDE.

Streak green, red, brown, white, grey. Hardness = 1.0,—4.0. Sp. gr. = 3.9,—8.2. If metallic, is black. If not metallic, is adamantine. If streak green, the colour is iron-black. If streak brown or white, the sp. gr. is between 4.0 and 4.2; tessular, cleavage perfect. If streak red, the sp. gr. = 4.5, and more; hardness = 2.5, and less. If sp. gr. = 4.3 and more, the streak is red.

Genus I.—Manganese-Blende.

Glanz-Blende, *Mohs.*—Manganese-Blende, *Werner.*—Manganese Sulphuré, *Haüy.*

Tessular. Streak green. Hardness = 3.5,—4.0. Sp. gr. = 3.9,—4.05.

1. *Hexahedral Manganese Blende. Sulphuret of Manganese.*

Hexahedrische Glanz-Blende, *Mohs.*—Manganese Sulphuré, *Haüy.*

Specific Character.—Tessular. Cleavage hexahedral, perfect.

Description.—Colour iron-black, which, on exposure, becomes brownish-black. Streak greenish-grey. Occurs regularly crystallized; in granular concretions, massive, and disseminated. Lustre splendent, shining, and metallic, inclining to imperfect metallic. Slightly brittle. Opaque.

Constituent Parts.—Protoxide of manganese, 85.0, sulphur 15.0. *Vauquelin.*

Geognostic and Geographic Situations.—This rare mineral occurs in Cornwall and in Transylvania, in primitive and transition rocks.

Genus II.—Zinc Blende, or Garnet-Blende.

Tessular. Streak grey, white, reddish-brown. Hardness = 3.5,—4.0. Sp. gr. =4.0,—4.2.

1. *Dodecahedral Zinc-Blende.*

Dodecaedrishe Granat-Blende, *Mohs.*—Blende, *Werner.* —Zinc Sulphuré, *Haüy.*

Specific Character.—Tessular. Combination semi-tessular with inclined faces. Cleavage dodecahedral, very perfect.

Description.—Colours brown, yellow, grey, green, red, and black. Occurs regularly crystallized, also in granular and fibrous distinct concretions, massive, and disseminated. Lustre ranges from specular splendent to glimmering, and is adamantine. Ranges from transparent to opaque. Brittle, and easily frangible.

The varieties in which the yellow colours predominate are named *yellow zinc-blende;* those in which the brown colours predominate, *brown zinc-blende;* and, lastly, those in which black is the characteristic colour are named *black zinc-blende.*

Constituent Parts.—Zinc 65.50, iron 1.37, sulphur 32.63. *Thomson.*

Geognostic and Geographic Situations.—Beautiful yellow varieties are met with in the old lead-mines of Tyndrum, in Perthshire; the brown, or common blende, in every lead-mine in England and Scotland; while the black variety, which is the rarest, occurs in small quantity in Saxony, and some other mining countries on the Continent of Europe.

Genus III.—Antimony-Blende. Purple Blende. Red Antimony.

Purpur-Blende, *Mohs.*

Prismatic. Streak cherry-red. Hardness = 1.0,—1.5. Sp. gr. = 4.5,—4.6.

1. *Prismatic Antimony-Blende, or Purple Blende.*

Prismatische Purpur-Blende, *Mohs.*—Roth Spiesglaserz, *Werner.*—Antimoine Oxidé Sulphuré, *Haüy.*

Specific Character.—Prismatic. Combination hemiprismatic. Cleavage prismatoidal, very perfect.

Description.—Colour cherry-red, frequently with a tempered steel tarnish. Occurs regularly crystallized; in granular, scopiform, and stellular fibrous concretions. Lustre shining and adamantine. Opaque, or translucent on the edges.

Constituent Parts.—Antimony 67.5, oxygen 10.8, sulphur 19.7. *Klaproth.*

Geognostic and Geographic Situations.—Occurs in small quantity in primitive rocks, in Saxony, France, and Hungary.

Genus IV.—Ruby-Blende.

Rubin-Blende, *Mohs.*

Rhombohedral. Prismatic. Streak red. Hardness = 2,0,—2.5. Sp. gr. = 5.2,—8.2.

1. *Rhombohedral Ruby-Blende, or Red Silver. Ruby Silver.*

Rhomboedrische Rubin-Blende, *Mohs.*—Rothgiltigerz, *Werner.*—Argent Antimoiné Sulphuré, *Haüy.*

Specific Character.—Rhombohedral. Rhombohedron = 108° 18′. Combination sometimes with different planes on opposite extremities. Cleavage rhombohedral. Streak cochineal-red. Hardness = 2.0,—2.5. Sp. gr. = 5.4—5.9.

Description.—Colour between cochineal-red and dark lead-grey, and sometimes inclining to carmine-red. Occurs regularly crystallized; massive and disseminated. Lustre ranges from splendent to glimmering, and is adamantine. Fracture uneven or conchoidal. Ranges from opaque to transparent. Sectile.

Constituent Parts.—Silver 58.94, antimony 22.84, sulphur 16.61. *Bonsdorf.*

Geognostic and Geographic Situations.—In some silver mines it is an abundant ore; in this country it has been hitherto found only in Cornwall.

2. *Hemiprismatic Ruby-Blende. Miargyrite.*

Hemiprismatische Rubin-Blende, *Mohs.* — Rothgiltigerz, *Werner.* —Miargyrite, *Rose.*

Specific Character.—Hemiprismatic. $\frac{P}{2} = 128° \ 59'$. $P + \infty = 86° \ 4'$. Cleavage $\frac{\frac{5}{4}\breve{P}r + 2}{2}$, $\breve{P}r + \infty$, imperfect. Streak dark cherry-red. Hardness = 2.0,—2.5. Sp. gr. = 5.2,—5.4.

Description.—Colour iron-black. Streak dark cherry-red. Occurs in crystals more or less regularly formed. Lustre intermediate between metallic and adamantine. Fracture conchoidal and concealed foliated. Opaque, but when viewed by transmitted light, thin splinters shew a deep blood-red colour. Very sectile.

Constituent Parts.—Antimony 39.14, silver 36.40, sulphur 21.95, copper 1.06, iron 0.62. *Rose.*

Geognostic and Geographic Situations.—This very rare mineral occurs along with argentiferous arsenic pyrites, in one of the mines of Braunsdorf, near Freyberg in Saxony.

3. *Peritomous Ruby-Blende, or Cinnabar.*

Rubin-Blende, *Mohs.*—Mercure Sulphuré, *Haüy.*

Specific Character.—Rhombohedral. Rhombohedron

$= 71^\circ 47'$. Cleavage $R + \infty$, very perfect. Streak scarlet-red. Hardness = 2.0,—2.5. Sp. gr. = 6.7,—8.2.

Description.—Colours cochineal and scarlet-red, sometimes inclining to dark steel-grey. Occurs regularly crystallized; in granular and globular concretions; massive, disseminated, and dendritic. Lustre ranges from splendent to glimmering, and is adamantine, verging on semi-metallic. Fracture uneven, conchoidal, and earthy. Ranges from transparent to opaque. Sectile, and easily frangible.

The dark-coloured varieties, inclining to steel-grey, are named *hepatic cinnabar*, the others *common cinnabar*.

Constituent Parts.—Mercury 84.50, sulphur 14.75. *Klaproth.*

Geognostic and Geographic Situations.—Cinnabar is named "the ore of mercury," because the mercury of commerce is obtained from it. The principal mines of this mineral are Almaden in Spain, and Idria in Carinthia. It also occurs in the Palatinate, Carinthia, Transylvania: Japan, Mexico, and Brazil. The rocks in which it is contained are gneiss, porphyry, greywacke, limestone, and the coal formation.

Order XIII.—SULPHUR.

Not metallic. Colour yellow, red, brown. Prismatic. Hardness = 1.0,—2.5. Sp. gr. = 1.9,—3.6. If monotomous, the sp. gr. = 3.4, and more; if sp. gr. above 2.1, the streak is yellow or red.

Genus I.—Sulphur.

Prismatic. Hardness = 1.5,—2.5. Sp. gr. = 1.9,—3.6.

1. *Prismatoidal Sulphur. Orpiment.*

Prismatoidischer Schwefel, *Mohs.*—Gelbes-Rauch gelb, *Werner.*—Arsenic Sulfuré jaune, *Haüy.* Auripigment.

Specific Character.—Prismatic. $\bar{P}r = 83° 37$. $P + \infty = 117° 49'$. Cleavage $\breve{P}r + \infty$, perfect. Streak lemon-yellow. Hardness = 1.5,—2.0. Sp. gr. = 3.4,—3.6.

Description.—Colour lemon-yellow. Occurs regularly crystallized; in granular and concentric lamellar concretions; massive, disseminated, stalactitic, reniform, botryoidal, and in crusts. Lustre splendent, between adamantine and semi-metallic. Translucent. Sectile. Flexible, but not elastic.

Constituent Parts.—Arsenic 61.86, sulphur, 38.14. *Laugier.* Is a tri-sulphuret of arsenic, therefore containing three atoms of sulphur to one of arsenic.

Geognostic and Geographic Situations.—Occurs in veins in various metalliferous formations in primitive and secondary rocks in Hungary and Germany.

2. *Hemiprismatic Sulphur. Realgar.*

Hemiprismatischer Schwefel, *Mohs.* — Rothes-Rausch gelb, *Werner.*—Arsenic Sulphuré Rouge. *Haüy.*

Specific Character. — Hemiprismatic. $\frac{P}{2} = 130° 1'$. $P + \infty = 74° 30'$. Cleavage $-\frac{\breve{P}r}{2} = 66° 44'$, $P + \infty$, $\breve{P}r + \infty$, imperfect. (Fig. 36, 30, 28.) Streak orange-yellow...aurora-red. Hardness = 1.5,—20. Sp. gr. = 3.5, —3.6.

Description.—Colour aurora-red, inclining sometimes to orange-yellow. Occurs regularly crystallized; massive, disseminated, and in flakes. Lustre shining, and adamantine. Fracture uneven. Translucent and semi-transparent. Brittle, and easily frangible.

Constituent Parts.—Sulphur 30.43, arsenic 69.57. *Laugier.* Is a bi-sulphuret of arsenic, therefore containing two atoms of sulphur to one of arsenic.

Geognostic and Geographic Situations.—Occurs in veins in primitive rocks, also in transition, secondary, and volcanic rocks. Germany, Hungary, Bohemia, Ætna, and Vesuvius, are localities.

3. *Prismatic Sulphur, or Common Sulphur.*

Prismatischer Schwefel, *Mohs.*—Naturlicher Schwefel, *Werner.*—Soufre, *Haüy.*

Specific Character.—Prismatic. P = 106° 38; 84° 58′; 143° 17′. P + ∞ = 101° 59′. Cleavage P, P + ∞, imperfect. (Fig. 30.) Streak white or sulphur-yellow. Hardness = 1.5,—2.5. Sp. gr. = 1.9,—2.1.

Description.—Colours yellow, brown, and grey. Occurs regularly crystallized; in granular distinct concretions; massive, disseminated, stalactitic, vesicular, and corroded. Lustre ranges from shining to glimmering, and is between adamantine and resinous. Fracture uneven, splintery, or conchoidal. Translucent and transparent. Brittle, and easily frangible.

Geognostic and Geographic Situations.—Occurs most abundantly in secondary, tertiary, and volcanic rocks; is met with also in transition and primitive rocks. Important localities are Sicily, Italy, Spain, and the island of Iceland.

CLASS III.

Specific gravity under 1.8. If liquid, the smell is bituminous. If solid, is tasteless.

Order I.—RESIN.

Hardness = 0.0,—2.5. Sp. gr. = 0.7,—1.6. If sp. gr. = 1.2, and more, the streak is white or grey.

Genus I.—Mellite, or Honey-Stone.

Melichron-Harz, *Mohs.*

Pyramidal. Hardness = 2.0,—2.5. Sp. gr. = 1.4,—1.6.

1. *Pyramidal Mellite, or Honey-Stone.*

Pyramidales Melichron-Harz, *Mohs.*—Honigstein, *Werner.*—Mellite, *Haüy.*

Specific Character.—Pyramidal. Pyramid = 118° 4′; —93° 22′. Cleavage pyramidal, but imperfect.

Description.—Colours yellow or red. Occurs regularly crystallized and massive. Lustre shining or splendent, and vitreo-resinous. Fracture conchoidal. Semi-transparent, or translucent. Brittle.

Constituent Parts.—Alumina 14.5, mellitic acid 41.4, water 44.1. *Wöhler.*

Geognostic and Geographic Situations.—This rare and remarkable mineral has hitherto been found only at Artern in Thuringia, where it is associated with brown coal.

Genus II.—Mineral-Resin.

Erd-harz, *Mohs.*

Amorphous. Hardness = 0.0,—2.5. Sp. gr. = 0.8,—1.2.

1. *Yellow Mineral Resin, or Amber.*

Gelbes Erd-harz, *Mohs.*—Bernstein, *Werner.*—Succin, *Haüy.*

Specific Character.— Solid. Colour yellow....white. Streak white or grey. Hardness = 2.0,—2.5. Sp. gr. = 1.0,—1.1.

Description.—Colours yellow and white. Occurs massive and disseminated; often incloses insects, leaves, and other parts of vegetables, also corals, &c. Lustre ranges from splendent to glistening, and is resinous. Fracture conchoidal. Transparent and translucent. Brittle, and easily frangible.

Constituent Parts.—Carbon 70.68, hydrogen 11.62, oxygen 7.77. *Ure.*

Geognostic and Geographic Situations.—This beautiful mineral occurs imbedded in alluvial and tertiary formations; rarely in secondary rocks. The finest masses are those found in the low, flat, and alluvial countries on the shores of the Baltic. It is occasionally met with on the coasts of Scotland and England.

2. *Black Mineral-Resin. Naphtha. Petroleum. Bitumen, or Mineral-Pitch.*

Schwarzes Erd-harz, *Mohs.*—Erdöl, Erdpech, *Werner.* —Bitume, *Haüy.*

Specific Character.—Solid...liquid. Colour black, brown, yellow, grey. Streak black, brown, yellow, grey. Hardness = 0.0,—2.0. Sp. gr. 0.8,—1.2.

Description.—Colours white, grey, yellow, brown, and black. Occurs massive, disseminated, globular, reniform, stalagmitic, and liquid. Lustre resinous, and ranging from splendent to glimmering. Fracture earthy, conchoidal, and slaty. Ranges from transparent to opaque.

The *liquid varieties* are named *mineral oil;* the *solid varieties, mineral pitch, or bitumen.* The yellowish-white, yellowish-grey, and wine-yellow, *liquid*, transparent varieties are named *naphtha;* the blackish brown, *liquid*, and translucent or opaque varieties are named *petroleum;* the blackish-brown *solid* varieties, with earthy fracture, are described under the name *earthy mineral-pitch;* the pitch-black solid varieties, with splendent shining lustre and conchoidal fracture, are the *slaggy mineral-pitch, or asphaltum* of authors; and lastly, the brown, massive, curved slaty, solid and elastic varieties are named *elastic mineral-pitch, elaterite*, or *mineral caoutchouc.*

Constituent Parts.—1. *Naphtha.* Carbon 82.2, hydrogen 14.8, *Thomson.*—2. *Petroleum.* When naphtha is exposed to the air it becomes brown, thickens, and seems to pass into petroleum; this shews the close affinity of the two minerals.—3. *Mineral Pitch, or Bitumen.* When petroleum is exposed to the air it thickens, and forms a mineral pitch resembling asphaltum.—4. The *Elastic Mineral Pitch*, according to Henry, consists of carbon 52.25, hydrogen 7.49, oxygen 40.10, nitrogen 0.15.

Geognostic and Geographic Situations.—The naphtha and petroleum flow from rocks of limestone and of the coal formations. The finest varieties of the former are found on the shores of the Caspian, the latter occurs at St Catharine's, in the vicinity of Edinburgh, and in several other places in Scotland and England. The earthy mineral-pitch is a rare mineral, and has been hitherto found principally in the Hartz; the slaggy mineral pitch is met with imbedded in the limestone, ironstone, and sandstone of the coal formation in the middle district of Scotland; and the elastic mineral-pitch occurs chiefly in the lead-mine called Odin, to the north of Castletown in Derbyshire.

Order II.—COAL.

Streak brown, black. Hardness = 1.0,—2.5. Sp. gr. = 1.2,—1.6.

Genus I.—Mineral Coal.

Amorphous. Hardness = 1.0,—2.5. Sp. gr. = 1.2,—1.6.

1. *Bituminous Mineral-Coal.*

Harzige Stein-kohle, *Mohs.*—Braunkohle, and Schwartz-kohle, *Werner.*—Houille. Jayet, *Haüy.*

Specific Character.—Colours black and brown. Resinous lustre. Bituminous smell. Peat smell. Hardness = 1.0 —2.5. Sp. gr. = 1.2,—1.5.

Description.—Colours brown, black, and grey. Occurs massive, ligniform, and rarely in columnar concretions. Lustre ranges from splendent to glimmering, and is resinous. Fracture earthy, conchoidal, slaty, and uneven. Opaque. Streak shining. Sectile or brittle. Easily frangible.

Those varieties of bituminous coal, in which brown is the predominating colour, with feeble lustre, more or less of the woody texture or form, easily frangible, with peat smell, are named *brown-coal,* under which division is included the *fibrous, earthy, aluminous, conchoidal,* and *trapezoidal brown-coal* of authors. The varieties in which the black colour predominates, and the resinous lustre is considerable, which are harder than the brown varieties, and give out on burning a bituminous smell, are named *black-coal,* of which the following kinds are enumerated by mineralogists, viz.

slate-coal, cannel-coal, foliated-coal, coarse-coal, and *pitch-coal.*

Constituent Parts.—1. *Brown-Coal.* Carbon 77.10, oxygen 19.35, hydrogen 2.54, earthy parts 1.00. *Karsten.* —2. *Best Newcastle Coal.* Carbon 84.99, hydrogen 3.23, oxygen 11.78.

Geognostic and Geographic Situations.—The brown coal occurs principally along with rocks of the tertiary class; the coal of Bovey in England may serve as an example of it in this island; the black-coal never occurs in tertiary formations, always in beds in secondary rocks, and principally in that group or series of strata and beds named the coal formation. The principal coal-mines in Scotland and England are situated in the coal formation.

2. *Non-Bituminous Mineral-Coal. Glance-Coal. Blind Coal. Kilkenny Coal.*

Harzlose Steinkohle, *Mohs.*—Anthracite, *Haüy.*—Glanz-kohle, *Werner.*

Specific Character.—Colour black. Imperfect metallic lustre. No bituminous smell. Hardness = 2.0,—2.5. Sp. gr. = 1.3,—1.6.

Description.—Colour generally iron-black. Occurs in fibrous and columnar distinct concretions; massive, vesicular, and disseminated. Lustre ranges from splendent to glimmering, and is imperfect metallic or silky. Fracture conchoidal, uneven, and slaty. Opaque. Some varieties soil. The varieties with splendent lustre and conchoidal fracture are named *conchoidal glance-coal;* those with slaty structure *slaty glance-coal;* the columnar varieties *columnar glance-coal;* and the fibrous and soiling varieties *fibrous glance-coal* or *mineral charcoal.*

Constituent Parts.—1. *Kilkenny*, 97 per cent. of carbon. —2. *Rhode Island*, 94 per cent of carbon.

Geognostic and Geographic Situations.—Glance-coal, like black-coal, occurs in beds in the coal formation in the secondary class of rocks. Kilmarnock, Saltcoats, and Sanquhar, may be mentioned as localities; it also occurs in veins and imbedded masses in secondary trap, as in the Calton-Hill at Edinburgh. It differs from the other kinds of coal, by its occasional appearance among rocks of the primitive and transition classes.

INDEX.

INDEX.

PRINTED BY NEILL AND CO., OLD FISHMARKET,
EDINBURGH.

Fig. 27.

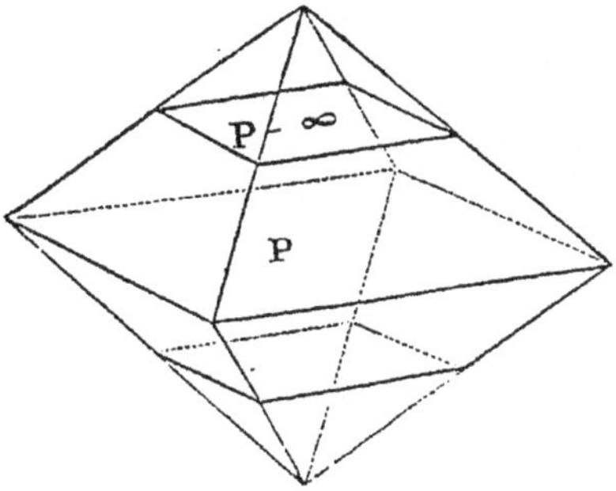

Fig. 28.

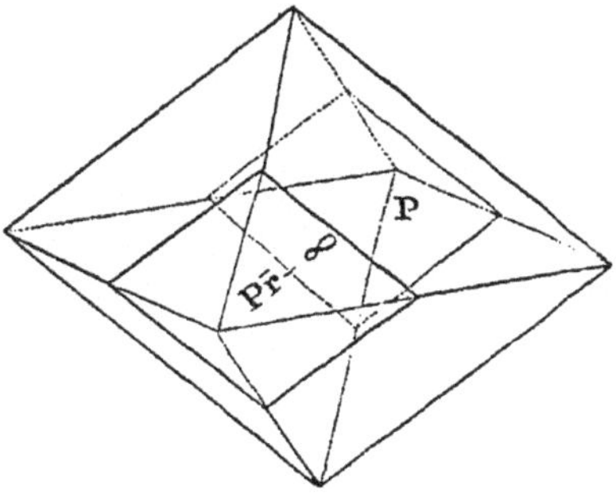

Fig. 29.

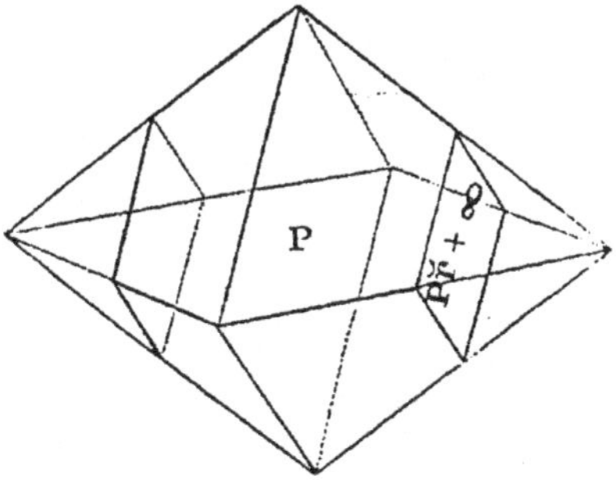

Fig. 30.

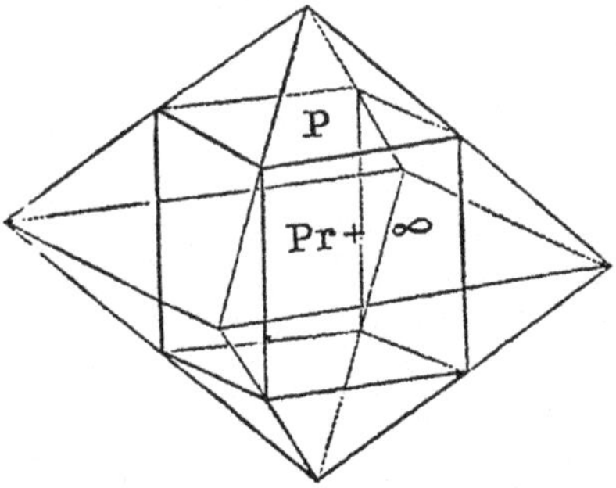

Fig. 31.

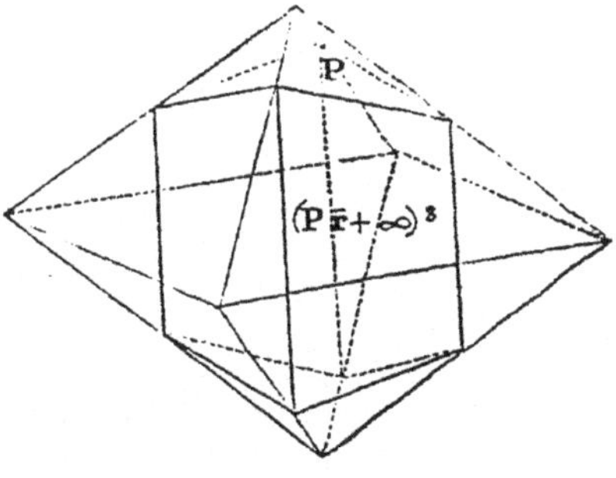

Fig. 32.

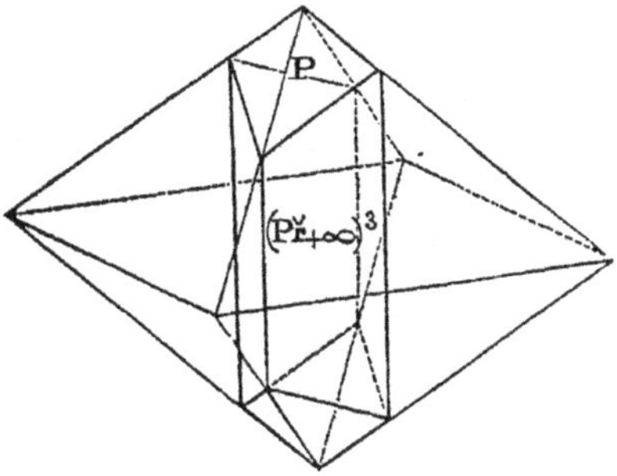

Fig. 33.

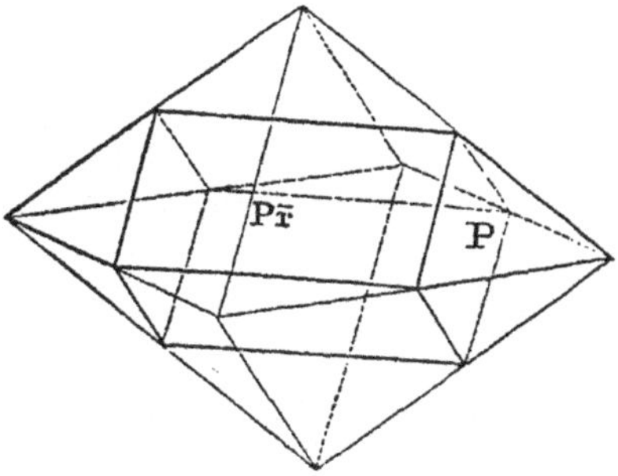

Fig. 34.

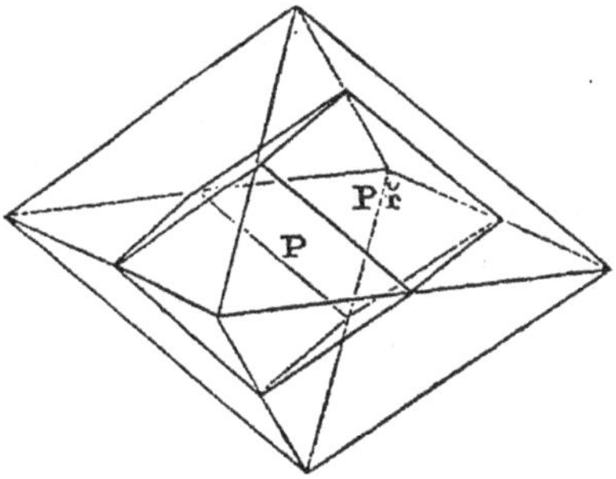

Fig. 35.

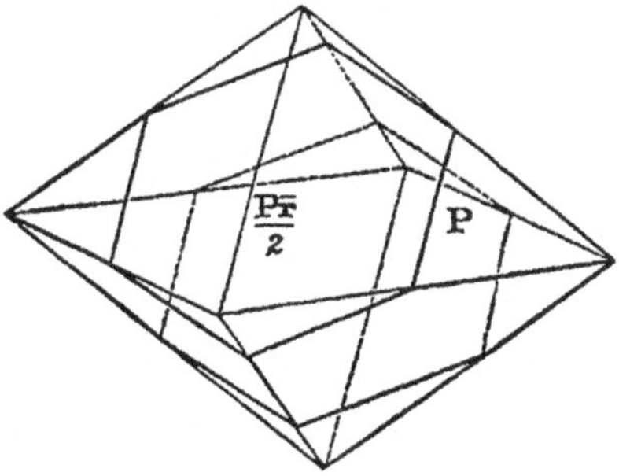

Fig. 36.

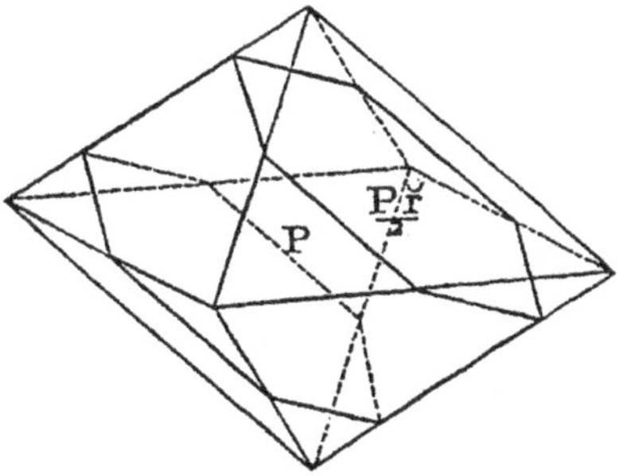

Fig. 37.

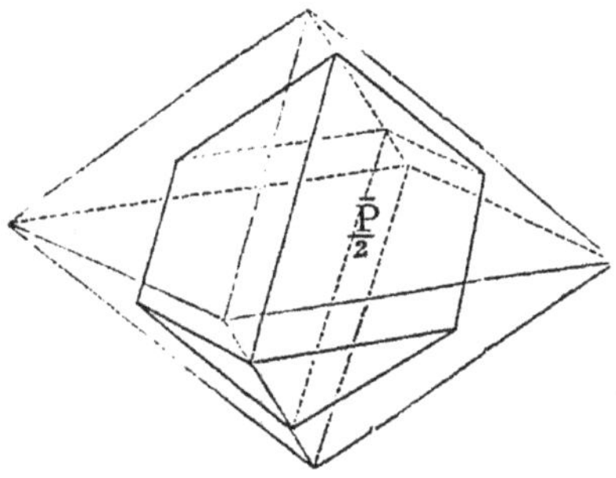

Fig. 38.

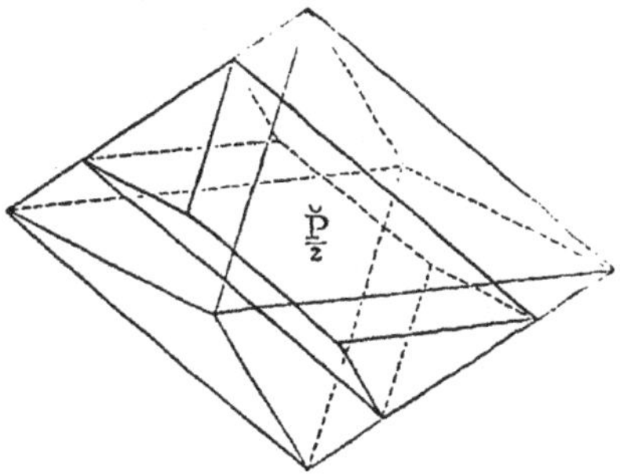

Fig. 39.

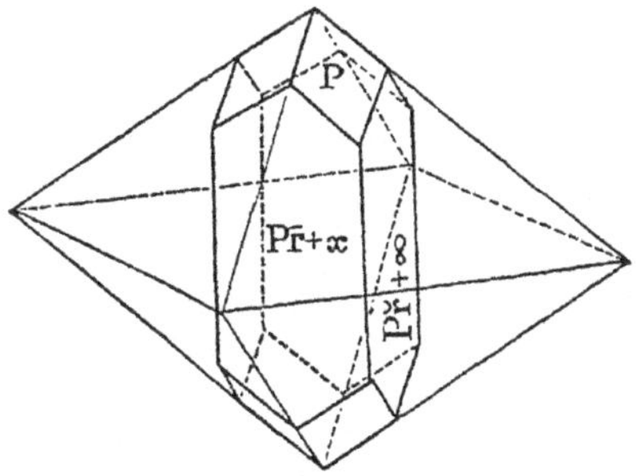

Fig. 40.

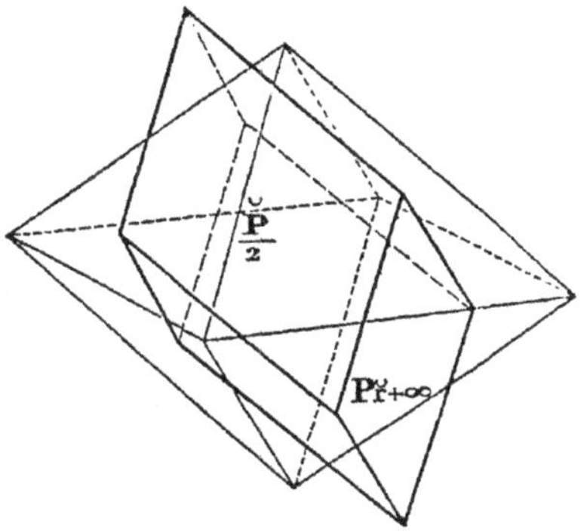

Fig. 41.

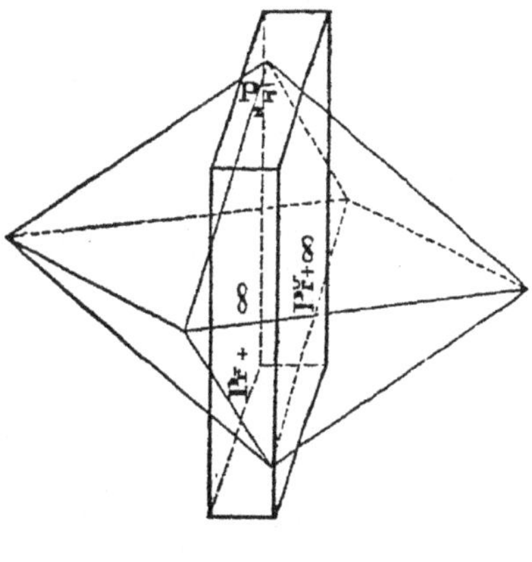

Fig. 42.

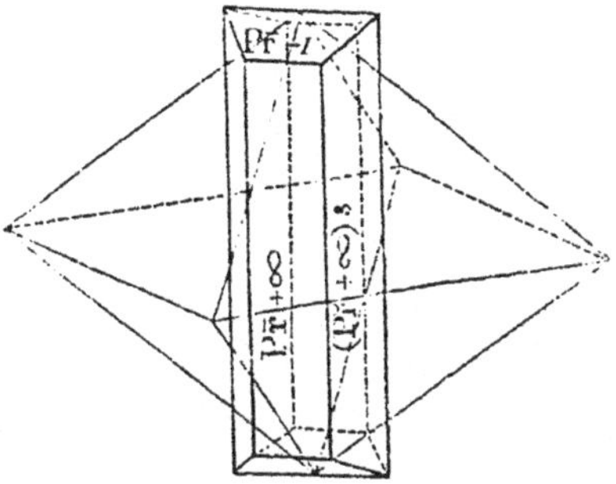

Fig. 43.

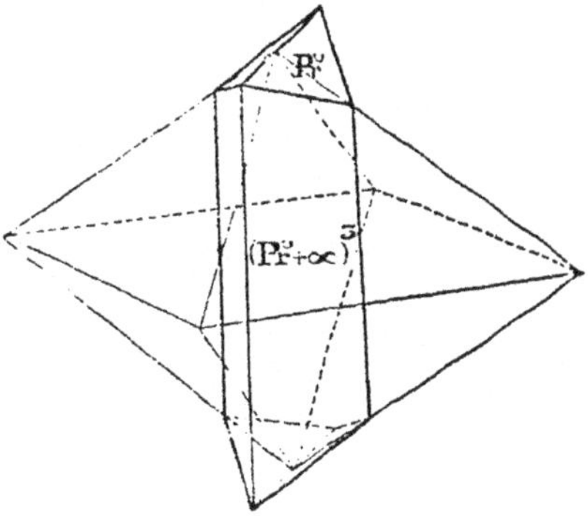

Fig. 44.

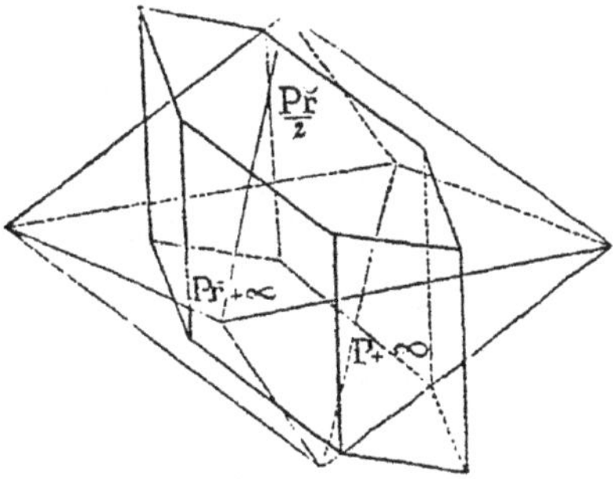

Fig. 45.

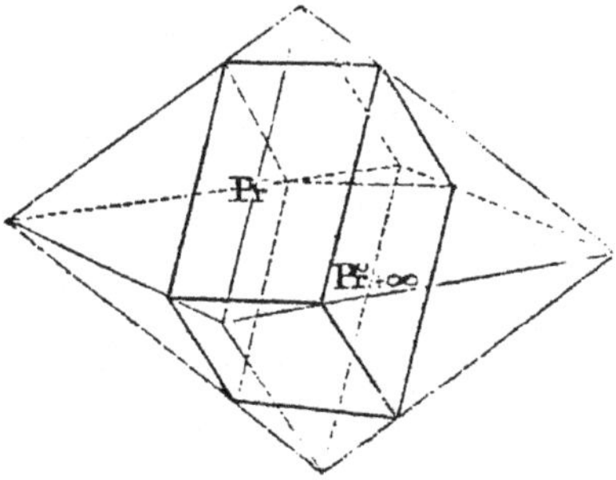

Fig. 46.

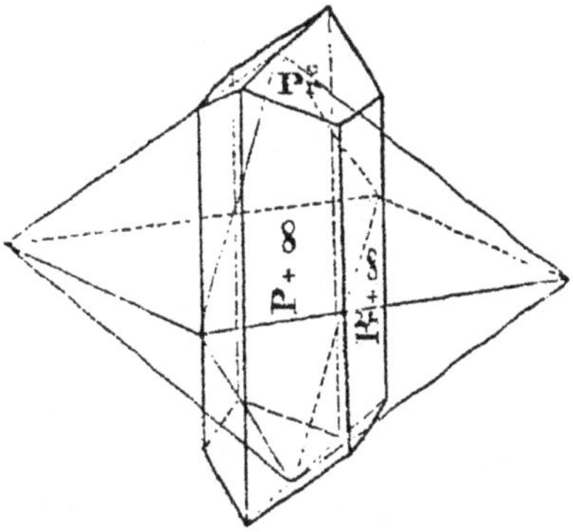

Made in the USA
Monee, IL
04 November 2026

67384988R10177